# The Future of the Automotive Industry

## The Disruptive Forces of AI, Data Analytics, and Digitization

Inma Martínez

Apress®

*The Future of the Automotive Industry: The Disruptive Forces of AI, Data Analytics, and Digitization*

Inma Martínez
London, UK

ISBN-13 (pbk): 978-1-4842-7025-7          ISBN-13 (electronic): 978-1-4842-7026-4
https://doi.org/10.1007/978-1-4842-7026-4

## Copyright © 2021 by Inma Martínez

Managing Director, Apress Media LLC: Welmoed Spahr
Acquisitions Editor: Natalie Pao
Development Editor: James Markham
Coordinating Editor: Jessica Vakili

Distributed to the book trade worldwide by Springer Science+Business Media New York, 233 Spring Street, 6th Floor, New York, NY 10013. Phone 1-800-SPRINGER, fax (201) 348-4505, e-mail orders-ny@springer-sbm.com, or visit www.springeronline.com. Apress Media, LLC is a California LLC and the sole member (owner) is Springer Science + Business Media Finance Inc (SSBM Finance Inc). SSBM Finance Inc is a **Delaware** corporation.

For information on translations, please e-mail booktranslations@springernature.com; for reprint, paperback, or audio rights, please e-mail bookpermissions@springernature.com.

Apress titles may be purchased in bulk for academic, corporate, or promotional use. eBook versions and licenses are also available for most titles. For more information, reference our Print and eBook Bulk Sales web page at http://www.apress.com/bulk-sales.

Any source code or other supplementary material referenced by the author in this book is available to readers on GitHub via the book's product page, located at www.apress.com/978-1-4842-7025-7. For more detailed information, please visit http://www.apress.com/source-code.

Printed on acid-free paper

# Table of Contents

# About the Author

**Inma Martinez** is an internationally acclaimed digital pioneer and A.I. scientist that advises business leaders and governments on how digital transformation can be leveraged to create competitive advantage and societal progress. For over 25 years she has worked across a variety of sectors - from user personalization, music and film streaming to intelligent apparel, smart cities and edge computing, driving forth their digital transformation. The automotive industry, where she has worked since the mid-2000s in vehicle connectivity and innovation as well as venturing out into the Formula 1 experience, is the sector that, in her view, will spearhead human progress towards a new digital civilization. In this book Inma shares detailed insights not only about how cars will be conceptualized, designed, manufactured, branded, and sold in the next 10 years, but also how their evolution into becoming digital machines has disrupted the core of this industry from transportation into transforming itself as a services sector that solves people's problems and addresses the green economy challenges.

Inma Martinez is currently a member of the expert group at the Global Partnership for Artificial Intelligence (GPAI), an OEDC and G7 initiative of worldwide reach with focus on innovation in the industrial and micro enterprise sectors.

www.inmamartinez.io
@inma_martinez

# About the Technical Reviewer

**Jenny Elfsberg** who currenty leads Sweden's government innovation agency, headed up the new innovation community workspace in Mountain View as part of Volvo Group Connected Solutions. The purpose with the hub is to become part of the Silicon Valley ecosystem and collaborate and co-create with different actors in the valley. During her time there, Jenny and her team focused on mobility, with a mission to explore and innovate with partners such as startups, tech giants, multinational corporations, universities, authorities, and decision makers.

Jenny has been with the Volvo group for 20 years, and in her most recent position prior to moving to Silicon Valley in August 2018 she was Director of Volvo Construction Equipment's (Volvo CE) Emerging Technologies organization since May 2009. Jenny and her global team of 20 research engineers worked to build knowledge and shape the future of the construction industry. The team created a future vision for the company, aiming at triple zeros (zero emissions, zero accidents, and zero unplanned stops) and also created the first generations of electric, autonomous, and connected concept machines. Jenny has also created a framework and a network with the mission to secure Volvo CE's innovative capability. Earlier in her career she has held a variety of senior positions, mainly within Engine R&D.

Jenny holds a Master of Science degree in Mechanical Engineering from Linköping Technical University in Sweden and a Licentiate of Engineering degree in Innovation Engineering from Blekinge Institute of Technology (BTH) in Sweden. Outside of work, Jenny enjoys many sorts of outdoor activities, learning new things, reading, and spending time with family.

# Foreword

When I first heard from Inma that she was writing this book, I was surprised by my strong reaction to this news. I was, of course, excited and wanted to read it, but I also felt a certain sadness, as this book would be a memorial of the car as we used to know it, and a book that would cover an era that will soon be in the past.

Inma and I share the same burning passion for technology advancement. We first met in 2018 at an industry event in London where I spoke about autonomous, electric, connected construction equipment, such as wheel loaders, excavators, and articulated haulers, and she spoke about artificial intelligence. It was wonderful to find a new friend, who is fun, energetic and warm, and who also shares my nerdy interest in technology advancements and applications, with both futuristic and historical perspectives.

I have been excited about electric cars ever since I sat behind the steering wheel in a tandem seated prototype of an electric drivetrain. This was maybe around 1996 or 1997 and it was developed by ABB Corporate Research for BMW. It was easy to fall in love with the silence, the acceleration, the immediate response when stepping on the gas. More than 10 years later, with my frequent business trips to Palo Alto in the San Francisco Bay Area, I followed the progress of how Elon Musk's Tesla cars were winning ground. First the Roadster, then my favorite - model S, which I was also able to see being made at the Tesla factory in Fremont, where they also manufactured the wire winding of their own electric motors. Then the X with the fun, but not so practical falcon wing doors, at least not in the snowy Nordic countries. Then the model 3 that many of my friends, including my dad, purchased and totally loved, and now the model Y, which is the model that finally is getting proper competition from automakers around the world. Even the team behind Ford's legendary Mustang is finally jumping on the EV bandwagon - with Mach 1.

During the first 10 years of my career I developed software in diesel engines for trucks, buses and construction equipment. Combustion engines are truly fascinating pieces of art. It is amazing to see how engineers keep building onto these over 100 year old inventions, making them better and cleaner, and optimizing them in every possible condition and duty cycle. Still, they keep utilizing liquid fuel and, no matter how many engineering hours organizations are investing in improving them, it is certain that electric machines will always be more efficient than internal combustion engines, where over 50% of the energy is lost to exhaust fumes and heat.

There are many challenges ahead before one could say that electric cars are always better for users, customers, the environment and human's health. There is the necessary infrastructure to be built and the system innovation that requires to optimize the entire value chain that was once needed for conventional cars, so let us be factual and fair when we transform our society into a more sustainable one. Many of us have loved the sound of an accelerating muscle-car, enjoying the luxury of driving up and down coastlines in convertibles. Still today many more people continue to rely on gasoline or diesel fuel to manage all their mobility wants and needs until the proper infrastructure is built across the land. We want to keep moving – but don't want to destroy the planet. This is why the future of electric vehicles is so interesting... Something other than the green economy has to ignite its hockey stick growth.

When it comes to self-driving vehicles, I feel like most people: fully aware of all the intelligent functionality already embedded in modern cars that makes us brake more safely, change lanes more collaboratively, and switch gears more efficiently. The fact is that many cars are already driving for us more than we understand. For most of us, these cars are appreciated as comfortable and convenient machines. Then, there are the carmakers that want to push for self-driving more, that give the driver the experience of letting go of not only the gas pedal, with cruise control, but also the steering wheel. All of us that have tried it know how profound that experience is. I love the convenience in busy traffic, where there is a lot going on in front and around you. I love it for when I drive across the country and want to be relaxed, but if I really want to get to know a car, I want it to be less smart and more manual, so that I can feel its unique characteristics.

When thinking about wicked problems in today's society, and the potential dangers that can be removed, of course the easily distracted, sometimes tired human driver behind the steering wheel of a fast moving and often heavy vehicle is a clear issue to tackle first. Surprisingly, it's really weird that, with all our focus on vehicular safety, we keep allowing humans to drive those machines. I'm not saying that the human driver should be immediately removed, but isn't it clear that traffic accidents should be avoided? Isn't it true that with more self-driving functionality we could get closer to zero accidents? I don't think that there is such a big step to fully autonomous cars if we just get the policies to catch up with technology. I'm sure this is a battle just as important as the exhaust emissions and clear air problems, but for some reason we seem to look at traffic accidents as tragic, but inevitable occurrences. Isn't it time that we reverse our mindsets and attitudes?

Surely, if we start to talk about policies that will make traffic safer, more convenient and more predictable, it is clear that we can forward the advancement of technologies that connect cars and infrastructure. The internet of things era has been around for more than 15 years and we all love our smartphones connectivity, the smart watches and other wearables, as well as kitchen appliances with WiFi which are the norm today. The connectivity of our cars is used for navigation but also to stream podcasts, and link up to conference meetings. Many carmakers collect this data from the cars to develop even better services. We also see promising new smart city solutions, where traffic lights are becoming more adaptive and coordinated and we've seen teams of big rigs (ie. trucks and trailers) connected to each other in what is called platooning. But there are so many untapped opportunities in this space – if we would utilize connectivity as a way to, not only collect data from different traffic system users, but also to orchestrate the traffic flow, hour by hour, weekday by weekday. It would enable cities to design the citizens' experience of living in an urban center, the truck drivers' productivity and the general safety and flow on the roads. These are promising initiatives going on, and as internet connectivity is evolving and becoming ever more stable, predictable and seamlessly adopted – basically taken for granted even, they have the power to make our modern society evolve into a new state of collaboration and coordination. Just like for electric vehicles and autonomous devices, this development will require not only technology and policy development but also a shared willingness by all parties involved to strive for the betterment of society and humanity.

I will always love the beauty of cars from the 60's and 70's. I will never get tired of hearing the sound of a powerful engine starting up and accelerating, and I'm childishly enjoying driving a fast car on winding roads. At the same time, I am hopeful and excited about everything that new technologies are enabling, and I appreciate that vehicles and roads are getting cleaner, smarter and safer thanks to our amazing ability to tackle problems and develop new solutions. For many years we have been able to develop for efficiency, convenience, profitability and growth. Now we have to take sustainability into account, and it is an urgent matter that cannot be dodged. I believe we can do this. We can utilize technologies and sustainable development goals to achieve a new era. We can celebrate cars as we have come to know them yet also lean into the new solutions. In this book Inma has beautifully connected the passion and love for cars and automotive industry with the interest and fascination for new technologies and new possibilities. We are not only witnessing a paradigm shift in the world, as mobility matters to every one of us and we are all part of it. We will probably see new modes of transportation emerge – modes that will make us look back at the car era in the way we look back at the horse carts of the nineteenth century – and say, perhaps, "Thank-God for innovators".

Jenny Elfsberg,
Stockholm, Sweden 12th April 2021
Head of Innovation at Vinnova, the innovation agency of the
Swedish government.
https://www.vinnova.se/en/
Formerly, Director of VOLVO Innovation Lab U.S.
in Mountain View, Silicon Valley.

# Acknowledgments

It has been a pleasure to write a book about this beloved industry because I have forged over the years a great passion for its innovation trajectory and the unparalleled talent of the many people involved in it, some of whom were not only willing and able to help me write this book, but also because they did it with delight, sharing the same admiration for its milestones and ability to create progress for humanity.

For that, I would like to thank my technical editor, Jenny Elfsberg, a "sister in arms" who has dedicated her entire career to develop innovation at Volvo Group at international scale, and who, like me, loves cars, driving them, and the end-to-end experience of automotive. I have spoken to many industry professionals over the years but I would also like to mention three people who opened my mind to many aspects of the automotive industry's evolution and prospects. My good-old friend from the early mobile Internet days, a veteran like me who has helped shape the digital world, Tom Raftery, Global VP, Futurist, & Innovation Evangelist at SAP who is a big advocate of sustainability practices and the green economy and who writes wonderful visions of the future on his personal website, visions that offer an unbiased view of what industries must do in order to advance humanity, not just industrial progress. I also loved talking to Steffen Hoppe, Director of Strategy at PwC Germany, who provided invaluable insights to the industry's business models and potential disruption of practices. With sincere gratitude, I am incredibly thankful to have had the opportunity to interview for the book Dr. Christian Dahlheim, Member of the Board of Management of Volkswagen Financial Services AG. He was both generous with his time and sincere in his views, which were shared with integrity and hope for the future, allowing me to confirm many of my assumptions for the sector, and to recognise how a heritage leader in this industry can rewrite history and disrupt it again and again and again. On that note, I must disclose that all my cars, the ones that I bought my own money, not hand-me-downs from family - my grandpa's Renault 8, or corporate cars - my 1995 Toyota MR2 from my banking days, have always come from this German brand: from my 1990 Golf GTI and my 2006 R32 to my current Audi 4 cabrio. *Vorsprung durch Technik*, the 1982 branding that Audi used across the whole of Europe in their campaigns - *in German!*, is the mantra of my tech and science career.

I have worked on and off in this industry since 2003. I am so grateful to have worked with engineers, designers, A.I. scientists, Human Factors experts and other talented, incredibly remarkable people who have responded to challenges and problems with brave dispositions. I can only hope that I have conveyed the greatness of their legacy with the degree of deference and admiration that they deserve.

Very special thanks go to Mark Gallagher, a Formula 1 executive with whom I had the pleasure of working with in the 2019 season and who has one of the most legendary historical memories of the sport. Thanks to him I was able to learn many of the intricacies of this five-ring circus of speed, technology, A.I. and passion. I am grateful for his mentorship and friendship. He has been able to transfer the big learnings and disciplines of this sport into executive education with his book *The Business of Winning - Strategic Success from the Formula One track to the Boardroom* (2014) and continues to produce valuable insights via his media collaborations.

There is also the inspirational book "How To Build A Car" written by the greatest Formula 1 vehicle designer Adrian Newey which my brother lent me. It is like reading Leonardo Da Vinci's autobiography and I recommend that you read it if you love all engineering aspects of the sport.

Every person involved in the development of this book, especially the talented editorial team at Apress Media New York, who have supported my writing and been wonderfully inspiring when suggesting additions and changes, has helped me realise my vision, which was to delight audiences and inspire them to appreciate the superb innovation delivered to humankind by this over a hundred-year-old industry.

I am passionate about the automotive industry because it has not only built innovations that have expanded beyond its remit and create joy and happiness for thousands of people, but also helped push forward progress and economic growth across the world, taking transportation to the next level, allowing us to send and receive goods, build infrastructure, get things done, and above all, keeping us safe from harm. Many people are excited about the future self-driving vehicles, and I am one of those people, too, but until then, I will watch with passion every new car model announcement, every Formula 1 season, every innovation and disruptive business model emerging within this sector because what happens in this industry is the stuff of dreams.

London, 8th April 2021

# Introduction

The idea of machines moving across the land by their own propulsion
has been forged across centuries and thousands of patents, engine wars,
aggressive branding, and consumer incentivized price-tags. Automobiles
stopped being utilitarian the minute we jumped aboard and felt the wind
brushing our hair while the forces of speed pushed us against the seats.
Nothing we owned gave us the same array of emotions, physical
reactions, and sense of future. Above all, cars made us independent,
one of the most coveted states of mind that humans seek along their life
paths. Still, unbeknownst to many consumers, automobiles have been
subject to the transformative forces of digital technologies ever since
electronics emerged in the 1960s as the next realm of life improvement.
Automobiles – cars, trucks, and other moving wheeled vehicles – were a
combination of technological advances and engineering paired up with a
kind prowess derived from passion, vision, and some pretty wild ideas that
turned out to work out for the best of everyone concerned.

To love cars is not about being a "petrolhead." It is to admire the
craftsmanship put into building, piece by piece, one of the most dynamic,
highly engineered machines ever conceived. Today, as cars become
luxurious contexts of delight – because nothing is more exhilarating than
sitting in a car on an open road and putting it to the test – in every sense
of the word, the automotive portfolio expands to all kinds of vehicles: tiny,
friendly, boxy cars like the ones they enjoy in Japan that zip up and down
the narrow streets and the rural roads; sedans, family station wagons,
SUVs, the cars that adorn our neighborhoods, that pull up in front of
schools to drop and pick up children; electric cars built by the big names in
the top tier of the industry, with minimalist designs and recycled materials

to encourage sustainability; campervans, people carriers, winnebagos, motorhomes crossing the wide open roads at sunset; excavators, bulldozers, diggers – the dream of every child that has ever played on a sandpit; trucks, tankers, eighteen-wheelers, the mighty fleets that cross the land like chariots of fire; planetary rovers, sturdy, resilient, curious explorers of Mars and the Moon. To continue innovating and optimizing this incredible invention is not just a craft, it is an ode to humanity's desire to prosper and better our lives.

When you are lucky enough to work in an industry where absolutely every person involved has this love of the craft, it is a joy and an honor. I have been lucky to have walked the big offices, with the big bosses showing me around, and the design desks, the factory plants, the small workshops, even the pit lanes at the Formula 1 where the garage floors are buffed so shiny and pristine that not even a speck of dust can be seen anywhere. I have a funny and heart-warming anecdote that I will share with you: I often wondered what makes a car, a luxury car, a high-end car, command such exorbitant prices. I was in Southern Germany, working on an innovation project at one of the classic, top automobile companies. I was learning what makes a top car become such coveted investment. I knew that if I would have asked the question to the people in suits, they would have given me all kinds of market data explanations. I was not interested in that. I have always loved to walk down to the yards, the garages, the pits, the places where the real, knowledgeable artisans build the most precious things. I went down to the factory plant, past the robots, and the wind tunnel. A magnificent, brand new, top of the line car was just finished being put together. Shiny. Very sparkly. The stitched leather, soft and creamy. The head of quality control was there, at the end of his shift, admiring this thing of beauty. I asked him: "Why is this car worth tens of thousands of euros?" He turned around and gave me a cheeky smile: "Get inside." We both did. "Now close the door." I did. Whoosh. The air inside the cabin was compressed, as if we had just closed the hatch inside a space capsule. The noises from the factory plant were muffled. We looked at each other "in the know." "Because of that," he said smiling. "Because it takes thousands of hours to get that door to close like that," as if it was on hydraulics. All of a sudden, we were on another planet, away from the mundane world. I knew he was right because at Nokia I had also been obsessed about what bearings should be used to build those slick, flashy cell phones that would slide open like in the Matrix, about how sturdy the buttons and levers had to be when pressing them down to give the user a sense that this gadget had been built with noble materials, not cheap plastics. "It takes a door to tell," I replied. "The engine is the easy part," he joked. "Can I take it home?" I teased him. "Not today."

Here is to the automotive family, which includes not just those who design and build amazing vehicles, but others who came along the way to innovate them, transform them, make them the stuff of dreams. And to you, who consider automobiles part of your household, of who you are, of what makes you feel safe, joyful, successful, optimistic, relaxed, futuristic. Let us reveal all the things that make them such wondrous things.

# PART I

# Car 4.0

CHAPTER 1

# OK Computer

*Mechanical engineering ignited the design concept of motorized vehicles, built for the purpose of supplanting horses and carriages. Devising different types of power unit propulsion systems, electronic engineers soon designed specific innovations that incrementally transformed vehicles into electrified systems and thereafter into computerized systems that automated many functional features of automobiles. Before vehicles were programmed to self-drive, they were digitally optimized devices.*

## Vehicle Electronics

Cars in the mid-twentieth century began to be more electronic than mechanical thanks to collaborations between electronic companies and car manufacturers. How cars became computerized systems is mainly thanks to a German electronics and robotics manufacturer called Bosch and a young physicist, Dr Heinrich Knapp, who in 1959 busied himself to devise schematics for an electronic gasoline injection system. Just like Intel became the *de facto* core processor of the majority of computers in the twentieth century, Bosch came to solve the ignition problems of petrol/diesel engines by computerizing the first electronic fuel injection models of the 1960s cars. The team of engineers assigned to the task had to resolve the challenges with ingenuity and entrepreneurial spirit. In order to understand what the real issues were, they convinced their senior management at Bosch to buy a Mercedes-Benz 300 car which

© Inma Martínez 2021
I. Martínez, *The Future of the Automotive Industry*,
https://doi.org/10.1007/978-1-4842-7026-4_1

they turned into their testing vehicle. In order to keep their research and development activities under wraps, they retrofitted the car back to its original carburetor injection every time they had to take it for mechanical servicing. They feared that Mercedes would suspect what Bosch was up to and reverse-engineer their prototypes and design approaches prematurely. The aim was always to show a car manufacturer their invention, but too soon would have diminished their ability to turn this into a commercial opportunity for the firm, who was gradually honing a great electronics know-how capable of providing solutions into one of the most coveted industries in Germany and the world.

*Jetronic*, Bosch's first electronic fuel injection system, was based on the rudimentary and unworkable but, nevertheless, pioneering technology developed by Bendix Corporation, an American manufacturing and engineering company supplier to General Motors and other automobile manufacturers. Bendix's *Electrojector* was a first attempt to produce an electronic fuel injection system for mass-produced cars, but it turned out to be troublesome and unproven. Jetronic's market timing, born some ten years after, benefited from the early development of computerized systems of the 1960s, and this time, its functional approaches did work. Still, selling to the motor industry the merits of electronically controlled technology took considerable evangelizing to convince the skeptics at some European and Asian car manufacturers. Bosch had an undeniable reputation selling other electrical mechanics to them, such as hydraulics and pneumatics, from power-window units to ignition in mopeds, but getting into the soul of a car, its engine, was another matter as this was an environment where a lot of things could go awfully wrong.

In real terms, the proud mechanical engineering teams at prestige car manufacturers were being told that a supplier of peripheral components was coming through the door to tinker with their precious engines. Disruption tends to encounter disbelief when sectors imagine themselves perennially unchallenged. In 1998 London I used to be brought to meetings with banks' CIOs to explain the merits of Internet banking.

I was usually rebuffed with incredulity by executives that sang the praises of bank branches for their luxurious decors or their value as community strongholds.

I remember a meeting with the chairman of Midland bank, a merchant bank that was later bought by HSBC. "Our clients love our branches!", he protested. After the acquisition, all those high-ceiling branches with cornices and big display windows onto iconic crossroads in London were purchased by a private equity firm who turned them all into a chain of bars, "All Bar One." Now their clients could well and truly love those branches, since they could hang out in them with a gin & tonic in hand. Digital disruption has sometimes a certain elegant irony to its devastation of old society pillars.

The reluctance of car manufacturers to adopt electronic approaches was forced to give in by the emergence of regulation regarding air pollution. Smog used to clog the air in car-inundated cities like Los Angeles, requiring the United States to become the first country to recognize the need for federal laws to resolve this health hazard. Between 1963 and 1968, stricter laws for air quality forced car manufacturers to consider other avenues to reduce their emissions. Volkswagen, riding the wave of success that their Beetle type 1 had achieved in the United States, was the target that Bosch approached to pitch their innovative injection system. Volkswagen was already manufacturing a type 3 Beetle which would never pass the new Air Quality Act regulation, so they had to put their pride aside and listen to what the electronic geeks had to share with them.

The success of the 1967 Volkswagen 1600 model later drove Mercedes and BMW to co-develop similar systems with Bosch. Years later, the 1976 Cadillac Seville, 1979 Toyota Supra, and 1981 Chrysler Imperial were models that came to market with this paradigm-shifting technology. Cars stopped being fitted with carburetors, and instead came with their own sensors to measure airflow and air temperature and a computerized system that could analyze this data and adjust the right amount of fuel that should be delivered to the engine. As early as 1967, a computerized system was, for the first time, directly responsible for the power, fuel economy, and lower emissions management of a motor vehicle. The years that followed the innovation of the original electronic injection, the technology continued to be prototyped and optimized according to tighter pollution regulations that began to spread worldwide.

---

Bosch is today one of the leading IoT companies in the world. The success of their Jetronic system not only helped ease the resistance to electronic components in car manufacturing, but provided the much-needed confidence to pave the way for further digitalization and computerization of car systems, from ABS brakes, airbag controls, and sensor-based parking assistants. As a heritage, trusted brand since 1889, Bosch has also developed white goods appliances under its own brand promise of building products for life, but in the last two years, it has unveiled an emerging vertical in IoT and robotics that spans back to their original ground-breaking roots of the 1960s, bringing to the present their entrepreneurial passion and bravado. "Living like a boss," that is, in control of our surroundings and the tagline to their CES 2019 innovation campaign, presents a fresh perspective on the growing power of automated electronics for personal use that are populating consumer shopping lists today. Domotics, that is, home automation, is now seamlessly expanding toward our nearby vehicles, who in turn, have become spaces that we inhabit with comfort, and almost equal gratuities as the ones we experience on our home sofa. This new reality, coming to us in 2021, has been a long-time coming and it plans to make your car an extension of your home habitat.

---

# The Internet of Vehicles: Car Sensors

Reliance upon electronics and sensors continued to plough through car engineering in the 1980s in parallel with the golden era of computer innovation. In 1987 microprocessors and the introduction of hard disk drives (HDDs) with self-testing functionalities – the baby steps toward machine automation – gained huge market shares for their manufacturers, British ARM RIC and Corner Peripherals, a San Jose, California company that had the good sight to partner with Colorado-based CoData and convince Compaq to finance their venture and act as the big distribution muscle for their product. The same year saw IBM shipping a million units of their first personal computer to be powered by an Intel processor chip, the 80386, a new operating system, OS/2, and the first computer mouse and 3.5-inch floppy disk. In Japan, Mitsubishi launched their first commercially available industrial robot, the Movemaster RM-501 Gripper, a sturdy little arm capable of assembling products or handling dangerous chemicals in labs, while in Europe, a consortium of telecommunications and handset manufacturing companies got approval from the European Union to deploy the first digital mobile network, the Global System for Mobile communications (GSM), a way to provide seamless connectivity as cell phone signals moved from cell to cell in the grid. GSM, an industry standard also adopted in Asia, was technologically superior to the existing CDMA networks in the United States. Only GSM networks were able to support the transmission of data and voice at the same time, a feature that, alongside SIM card portability, put AT&T and Verizon to shame when compared with the versatility of operations that Qualcomm was able to offer to US consumers. Before the official release of the Internet in April 1993, mobile networks were in advanced development, creating the precursory work that decades later proved to be the biggest digital transformation: mobile Internet. It should not surprise anyone then that at car manufacturing plants and design desks, electronics were now a prime tool for innovation. Digitalization was taking over the world, and the automotive world was soon to be an industry that would embrace its many challenges in return for the optimization of proprietary technologies and the competitive advantages that it created when marketing car models to consumers.

That very same year, General Motors improved exponentially the ignition system of their new models via sensors that provided orientation of individual ignition coils positional data. The clever piece of software handling this sensor data was then able to fire the individual coils connected to the ignition plugs with exact precision, creating what the entire automotive industry adopted as a *distributorless* ignition system or DIS. With one stroke of digital genius, the American car giant put an end to the traditional practice of connecting a single wire distributor to the ignition coil to raise the voltage and spark the plugs. An internal rotor would distribute electricity to each spark plug wire as it turned and the whole operation was kept going by simple mechanics. Until then, anyone with a head for mechanics and a good wrench could fix the moving parts of their own car. While electronic ignitions delivered additional values such as lower emissions, spark plug life to over 50,000 miles and made engines easier to cold-start, it also removed our ability to fix them, or even know what was wrong with them. This capability was now bestowed upon computer diagnostic machines. The DIS showed car manufacturers that sensors and software were cost effective and intelligent systems worth deploying across the wide spectrum of car parts, not just ignition systems. This was the firestarter moment toward turning cars into computerized systems. No one looked back. It was the future.

The growing complexity of engines drove the car industry to build what is referred to as On-Board Diagnostics (OBD) and to program engine systems to self-diagnose and report faults automatically. When they first appeared in the market, with the rise of fuel-injection engines, their feature landscape was quite basic. A light would appear on the screen unit but it would not provide any further clues as to what was wrong. OBDs were initially built to guarantee quality control at the assembly line point, and not really as consumer features, since drivers obviously could not fix anything by themselves at this point. The "idiot lights" – as they are referred to in the industry parlor – that eventually came to pop up in the car control unit were basically made to pinpoint that the car needed to be taken to a garage and be put in the hands of a qualified mechanic. In the United States, and again motivated by the pressure of the California Air Resources Board (CARB), by 1988 all new vehicles sold in the sunny state had to be fitted with diagnostics functionalities, and eight years later it became mandatory for all states. The European Union reacted to this measure in 2001, making it compulsory for gasoline engines, and by 2004 for diesel engines too, while Australia and New Zealand followed suit in 2006. By 2008 the United States adopted formal ISO signalling standard and, at manufacturer's discretion, every carmaker began to create their own proprietary systems to report on a variety of issues, that is, if the ignition was on, if the body and chassis were grounded (God forbid we would die electrocuted), and other data that would demonstrate the correct workings of all electrical systems, including battery voltage, fuel and air metering, emission controls, transmission, and computer output circuits.

Discreetly, our beloved cars were becoming increasingly more computerized and automatized. We did not raise an eyebrow at this because the marketing and branding departments of car manufacturers were steering our attention toward other value propositions such as performance, comfort, cool factor, and, for the first time, safety. Since the 1950s both car manufacturers and automotive enthusiasts came up with a variety of approaches to protect the lives of humans aboard. Seatbelts and airbags were tested, patented, and eventually adopted as official and mandatory safety features by the 1980s. The 1960s and 1970s options were cumbersome and impractical, causing shoulder and neck injuries and not inflating properly, but soon they would also be transformed by sensors. In 1981 Mercedes-Benz decided to integrate the functionalities of each into a single solution. On impact, airbags sensors would automatically reduce the tension in the seatbelts to allow for crash dynamics to perform less kinetic force on the human bodies strapped by them and allow the release of the safety bags to cushion the driver's body weight while blocking the dashboard. Initially a feature of luxury cars, automakers fitted only the driver seats as a necessity until Chrysler in 1988 was the first one to offer the driver-side airbags in all their US cars. By 1991 both passenger and driver airbags became mandatory for all vehicles.

# Fit for Purpose, Safe for Humans

Safety requirements also affected the innovation of headlights, which, in the previous decades, were a low common denominator car part comparable to batteries. The lights fitted to cars were utilitarian and not considered a proprietary asset of innovation or a product differentiator. One could just buy any kind and install them in pretty much any vehicle. When in 1992 European carmakers upgraded traditional halogen headlights to High Intensity Discharge (HID) ones, headlights stepped their game up and became a new field of car innovation, with HIDs being able to produce 3000 lumens (a measure of the total quantity of visible light emitted by a source per unit of time) and 90 candela per $m^2$ of luminous intensity superior to that of halogen lamps. With a higher degree of visibility, HIDs assisted drivers in seeing better in the dark, and therefore, driving with improved vision and increased safety.

The widespread use of sensors began to transform almost every car component into a source of informational data, and soon, a car's computational abilities to make sense of its environment and to react to it exponentially increased. A modern car does not just have a central computer, but various highly specialized computer systems that monitor and manage a wide spectrum of tasks. How carmakers began to factor in the need for computerized systems began with the realization that with faster speeds, car handling was in dire need of assistance. The car industry turned to software to resolve this problem, this time, consciously and with real purpose.

Electronic Stability Control (ESC), first developed by BMW in the late 1980s, was developed under the assumption that almost every driver inevitably panics and tends to untimely apply the brakes when losing stability, something that professional pilots and drivers on icy roads know it is best handled with the steering. Most drivers actually make things worse by braking and end up turning the car or sliding off the road. Co-developed with (again!) Bosch and Continental-Teves, a tires maker who has also evolved to become specialized in brake systems and other strategic car technical components, ESC was fitted to all of BMW car models of the year, including their new pride and joy the BMW Series 3, a low-cost model aiming to enter the high-volume mass market. The genius in this sensor technology resided in the predictive abilities of algorithms to advance when a vehicle was about to lose control, and – a giant leap in automation – intervene to maintain stability. This was one of the earliest pioneering milestones that is now leading the Vehicle Situational Awareness (VSA) developments: when a car has been programmed with AI to take over the driver's abilities in order to prevent an accident. In the ESC programming scenarios, software algorithms were trained to keep track of sensor data feeds from ABS brakes, steering wheel swerving, rotation rate of the car's trajectory – measuring by which angle the car would drive *off piste*, and turning force – the chances of the car spinning on itself doing what the Formula 1 calls "doughnuts." Soon, other competitors like Mercedes, Toyota, and Ford began to develop their own versions of ESCs, with more or less success.

The complexity of designing an algorithmic model for an ESC system is still one of the most concurred fields of post-doctoral theses for mechanical engineers, who continue to present novel approaches to the simulation programs used in the design of car systems. In order to teach car computers to advance and react to centrifugal forces, operate suitable maneuvers, and automatically stabilize the car, the data models need to comply with the design models conceived by engineers. CarSim and Matlab-Simulink of MathWorks have provided automatically generated code that renders the design models seamlessly without the need to know how to write real code in C and HDL. Regardless, the data complexity is still the biggest pickle to solve. An algorithm is only as good as the richness – the amount and variety of data sets that it is fed – and, above all this, the assurance that the data is true and not biased. In addition, stabilizing an object in movement, propelled by its own forces and subject to external forces that in turn operate on it, is one of the most challenging environments one can face. In fact, when one sets the task at "stabilizing the car" this is actually the compounding of tens of other stabilization modes that layer one upon the other forming a particular type of "cake from Hell" scenario for every engineer brave enough to take up the challenge of optimizing this feature. An engineer trying to resolve this issue will have to take into account two types of physics forces: Yaw Stability – when your vehicle spins on itself, and Roll Stability – when your vehicle is pushed sideways and rolls over when trying to steer a corner. This is one of the typical points of pilot feedback in Formula 1 when they discuss understeering and oversteering dynamics with their mechanics and aerodynamics teams. When the track turns, you want to turn the car through the corner. All very logical and sensible. But because it is a race, you want to do this as fast as possible. Here is where the car faces the most challenging gravitational forces. The tires, the only part of the car that makes contact with the track, obey to different dynamics. Rear tires are the ones that receive the output power of the engine, which is in the back in Formula 1 cars. The engine pushes backward and the tires push forward. The front tires, when pushed onward, present a certain resistance and are the most susceptible to sideways forces. When pushed by the rear tires, the front wheels will not go straight but instead, they will trace a curve. At slow speed, the curve will stay within the track. It is not the most effective way to race through a curve at high speed, though. This is why sports drivers will try to cut through a straight line toward the inner axis of the curve in order to bypass the curve as fast as possible rather than going all around it. In order to turn to the left, for example, the front tires will have to grip the track while resisting the inertia of still going straight. If the forces are too strong, you will see in races how the left front tire will lift off the ground, twisting the car aerodynamics, and the car will not be able to get closer to the inner axis of the curve, sliding off at the outer edges. This is understeering. The car's front wheels do not have enough grip and cannot turn into the curve properly. Oversteering has to do with how the back tires handle the curve. The mass of the car affects the back tires because the actual power of the engine is also pushing the car to continue going straight. If the speed is faster than the back tires can handle, the car back will slide off the ideal tracing of the curve. If you want to see how this becomes the greatest show on Earth, watch the Monaco Grand Prix, a city track where the angular curves of the streets put all racing cars to the absolute limits. Formula 1 drivers push their cars to over 200 mph and sustain gravitational forces of 5G and 6G in the curves. The fastest speed in Formula 1 tends to be achieved right after a long straight and before the breaking zone. Breaking late is what all drivers try to do, and they manage to keep their cars gripped to the ground if the aerodynamics are absolutely nailed in the design of its chassis. Driver and car are one machine that must push its bearings with every steer, acceleration and brake. Transferring this engineering toward their factory cars is what all constructors aim to achieve. This is why Formula 1 is a playing field for sensors, data feeds, algorithms, as well as aerodynamics, mechanics, and pushing the entire car to its limits. If there is a car that resembles a jet fighter, it is a Formula 1 vehicle. And thanks to digital technologies, the teams are supported by data scientists as well as mechanical and aerodynamics engineers. The car is a computer that flies on the track enduring Mach 1 speed forces. It is one of the most exhilarating spectacles to watch live.

# Automation

The decision to build "intelligent cars" that would take over the driving was a significant departing point for some manufacturers. For Mercedes Benz, working alongside the inescapable Bosch, the aim was to target a new car design paradigm within the luxury car market: less driving and more comfort. This decision diverged from the path that BMW, its homebrew competitor, was pursuing at the time. BMW's spunky, high-energy model, the BMW Series 3, aimed at a completely opposite consumer: a driver that loves to experience the handling of a fast, sporty car. The German Autobahns, authorized to be zero speed limit highways, became racing tracks where the young at heart and the cool drove *Beemers* and the old-fashioned and the conservatives sat aboard their expensive *Benzos*. While Toyota developed its own version of ESC which they called Vehicle Stability Control System (VSC), designed in collaboration with their own Asian component partners, Aisin and Denso, Ford was busy developing its first production system for the 1999 Lincoln LS with supplier ITT (now Continental-Teves) but also delving into one of the most successful consumer verticals in car manufacturing: SUVs, a type of vehicle that has redefined American driving. Many industry commentators have wondered why American car manufacturers came late to the ESC party. Their decision was purely commercial. Americans wanted to drive big-muscled pickup trucks and SUVs. Designing ESCs for these high-reaching, heavy vehicles was problematic because they stood at a much higher height from the ground than normal cars. This made the dynamic forces of rotation and turning a nightmare to control because the farther away that the car chassis is from the ground, the higher the probability of cars turning over themselves when these forces reach high speed and the tires' grip is compromised. ESCs became a technical development for luxury cars as opposed to being a fixture of urban tractors, and "fat and mighty," oversized vehicles. The same year, in the United States, GM also equipped their 1997 Cadillac DeVille, Eldorado, and Seville with an ESC design called *Stabilitrak* made by Delphi Technologies, a British manufacturer of fuel delivery, power management, and vehicle service needs solutions. Later adapted to a number of other GM vehicles, the innovation around this automated technology was carefully enunciated as a feature allowing drivers to drive "better," eliminating any threat to their own abilities being undermined or suppressed by an automated system that was engineered to take over.

ESC began to be understood and developed as an automated system that would take over the driving if the algorithm came to realize that the human at the wheel would handle the vehicle worse than what the engineers had already factored in the data hypotheses and assumptions. This important milestone was a key defining element for the likes of Mercedes Benz, who included this new technical wonder in their marketing materials and advertisement of 1997, describing ESC as a functionality that could stabilize the car electronically and continue to be "on call" if the driver was to lose control. For the first time, consumers were let known that their cars were actually smart and capable enough to take over in order to prevent an accident. Car electronics, with the subtle move of making cars safer and easier to drive, actually achieved something more visionary: to remove the driver from the equation, even if for a fleeting moment in time. In those split seconds, a machine would drive itself the best way it had been programmed to drive and everyone seemed to be fine with it. In fact, they were delighted at the thought of it.

What makes this moment fascinating to me, an artificial intelligence pioneer who, since the 2010s, has been challenged to explain and defend the development of machine intelligence in the future society, is that no one has ever associated AI in car automation to a picture of terminator machines that would rise up and, one day, "take over" in order to exterminate the human race. As far as ESCs go, cars have been taking over for decades now, and with the prospect of one day being driven by a self-driving car on the horizon, humans are okay with this, all thanks to branding and the realization that fast vehicles require even faster reactions. Sentiently, our cars began taking over the driving way before Google started to test their self-driving cars in their Mountain View car park.

Engine management sensors expanded to other versatile deployments. While the mid-1990s saw the development of the much-needed and appreciated on-board diagnostics – knowing when you are about to run out of petrol or if there is a failure in some other component – and other advancements in engine performance and petrol consumption, Mercedes-Benz continued to up their game by bringing consumers a flavor of what cars would become in the future: environments where the delight of consumer electronics would dash audiences. The Smart Key, launched by Mercedes Benz in 1998, was the stuff of sci-fi movies: car keys became fobs that would be detected by sensors in the door-locking and ignition systems as one would approach the vehicle. Starting a car became the push of a button, and fumbling for keys inside your pockets became obsolete as the car fob, a small pebble in the ocean of items that most of us carry in our handbags, transformed the car experience into a valet service. Your car was "recognizing" you on approach and was elegantly ready to go.

This commercial decision has turned its rewards for Mercedes-Benz. According to a yearly study of the global automotive technology market *"From Mercedes to McLaren: The most innovative carmakers"* conducted by the Center for Automotive Management, an independent market research and consulting institute near Cologne, Germany, in 2019 Mercedes-Benz jumped ahead in the innovation charts with around 180 innovations in a series production of which 77 are considered "absolute innovations" or world firsts. Mercedes-Benz, leading the table with 390 index points over its nearest competitor, Audi, demonstrates the long-term vision that the carmaker set for themselves since the 1990s: to deliver a brand focused on creating luxury experiences but above all a clear promise: "We want to keep you safe," a byline that you can still read on their current website.[1]

# Driving Forces of Car Innovation

*Safety* and *Better* became the driving forces of car innovations throughout the 2000s and the innovation targets of their digitization. Safety became such a pillar of car innovation that Swedish Volvo has strongly built a pristine reputation around safety features.[2] Even though for years their cars were still designed like Legos, square and boxy, yet simple and approachable, Volvo became throughout its long history the leader in creating customer trust with their crash dynamics tests and the protection of passengers, reliable and as grounded and humble as a *smorgasbord* food spread: buttered bread with toppings, nothing to hide, all out there in display, Swedish values that run deep into every component and design approach. The concept of "Better" presents different dynamics: it is about optimization, not innovation per se. It is about attempting perfection, iterating every aspect of design and engineering that can be

---

[1]How Mercedes-Benz has achieved such a footprint in the car market has its roots in the country's legacy of innovative engineering. Germany is a country of crafty engineers who thrive to build the best of everything. It is a national pride and something that speaks of their ability to push the limits in innovation. Today Mercedes, synonymous with "prestige bling," fends off both Audi, whose motto translates to "Innovation via technology," and BMW, a brand leading by their handling and fast performance, with solid sophisticated technological advances but not luxury. Furthermore, Mercedes-AMG Petronas' Formula 1 sports racing team has been winning the constructors' championship for six consecutive years since 2014. They are the contenders to beat and the biggest stand at Frankfurt's Car Expo every September. To be that obsessed with perfection has paid off.

[2]Vehicles were fitted with belts in the early days of motoring, but they often injured drivers and did more harm than good until Swedish engineer Nils Bohlin developed for Volvo the now traditional three-point seat belt, ergonomically adapted to wrap around a human body with better grip and less prone to cause injuries.

improved, thriving to stretch every part to its potentiality. Volvo is doing this precise move in its quest to offer the safest cars. It has conducted detailed tests on how women's bodies have never been taken into account when designing seatbelts as crash test dummies are built to resemble the bodies of men. So Volvo has launched the E.V.A. Initiative: to make their cars safer for women, sharing their 40 years plus of accident research with the rest of the automotive industry, a bold move taking an open source approach to automotive design. Still, there is a realm above the notion of "Better." It implies to disrupt a paradigm, challenge a mindset, break up a traditional approach to solving a problem in order to propose a solution that bats it out of the park in unexpected ways. This, and nothing else – accept no substitutes, is what innovation really means. Car engineering, as we have seen, soon embraced innovation as a way forward since its early decades in order to steer toward engineering and design paradigms where customer appreciated values such as safety, control, speed, and luxury, would power manufacturers' best efforts and aims by technological innovations where digitization of systems and sensor-based technologies played enormous roles. The soul of the automotive industry is audacious, knows no borders, and dreams up futures like no other consumer goods industry. Whether we drive fully electric cars or connected and automated vehicles in the next five years, the truth about this industry is one that dares to build the impossible, embeds itself deeper into our lives, and as such, becomes more and more digital, expansive, and all encompassing.

## Summary

Electronics are the core components of car innovations and allowed engineers to transform the driving experience into a safer and more efficient activity. Cars have iterated in design concepts and engineering toward both improvements and optimization of processes, components, and features, but increasingly so to become computerized intelligent systems, automated machines that have delivered to every prerequisite of both customers and motoring regulators, government authorities and the need to create competitive attributes to gain bigger market share. As soon as the forces of digitization began to transform business and society with computerization, as soon as cell phones became entertainment devices connected to the Internet, cars also pioneered the rise of digital dashboards and in-car infotainment. They were, let us say, the closest thing to our emotions, the spaces where we lived some of our most memorable moments, and, full of electronics, to become digital devices was a very straightforward step.

# CHAPTER 2

# Mission Control

*As vehicles began to cruise the roads, seamlessly flowing in traffic, our capacity to multitask inside their cabins began to pave the way to interiors and multimedia consoles favoring all kinds of activities other than paying attention to the roads. Cars began to become extensions of our offices, our homes, even concert halls, cinemas. Screens began to take considerable real-estate in the car consoles. We began to manage dashboards, menus, slowly letting the steering to our car's intelligent systems.*

## A Moveable Feast/Let Me Entertain You

When the car industry mastered the sensors, the algorithms and the transformation of cars into sophisticated "sedans," digital disruption was just around the corner. Though technology at the start of the millennium continued to become an embedded component into the car systems, a total game-changer arrived at the sacred passenger cabin that completely ignited the minds of most car manufacturers: *in-car infotainment.* The emergence of music and video content streaming and its evolution to smartphone platforms demanded of cars to become a continuation of our entertainment lifestyles. Cell phones, as networks expanded their data transmission capacities from 2G and 3G to 4G, became the ubiquitous gadget in our lives. Transferring the smartphone experience to automobiles required the installation of Bluetooth into vehicle consoles and with this the addition of a revolutionary

© Inma Martínez 2021
I. Martínez, *The Future of the Automotive Industry*,
https://doi.org/10.1007/978-1-4842-7026-4_2

satellite network: GPS navigation. Cars in the 2010s became interactive spaces allowing drivers to make phone calls over the stereo system and training the early adopters of the first voice commands systems. "Call home" is probably a sentence we have all said a thousand times while sitting in our cars more than when today we scream across the living room at a conical gadget called Alexa, Nest, or Siri. Realizing the potentiality of turning car interiors into contexts for human experiences, car manufacturers began to look into optimizing the head units by transforming them into touch screens similar to those in smartphones. By the mid-2000s, the concept of car infotainment expanded to the steering wheel, which was for the first time fitted with touch buttons to control the media menus without taking one's hands off the wheel.

When in 2007 Apple launched its first version of a smartphone, it basically turned their music player iPod into a cell phone with a web browser. The first version of iPhone was a very basic cell phone – it came without SMS, which puzzled and equally annoyed the European and Asian consumers, but it planted a flag into a highly prized peak: the digital entertainment market. iPhones completely took the mobile market by storm because they achieved what no other handset could before them: to become a true multifunctional cell phone, social and entertaining first, and a telecommunications device after. With it, thousands of people took their music and video libraries to the street. The rest of the handset manufacturers were rudely awakened from their perceived superiorities, achieved in the late 1990s and early 2000s. "On-the-go" became the de facto word for every single digital service aiming to attract consumers to their platforms. Google was pressing hard for their Pandora box–style G-platform, where a multitude of services beyond Gmail enticed users toward their Android operating system, and Tom-Tom, a popular navigation service, began to be deployed across both mobile and car navigation screens. The drumbeats soon got to the ears of the automobile makers, who had been building "moving" and "on-the-go" artefacts since the 1900s. Well aware that car buyers were now demanding multimedia services within the context of their driving experience, around 2006, informally at first, and then via regular meetups, some car manufacturers in the Bay Area, began to discuss approaches and opportunities "following the realization that the delivery of increased functionality in automotive infotainment solutions, particularly in the area of device and services connectivity, was becoming unsustainable."[1] Software was not cheap, and deciding on what platform to adopt made the idea of developing a proprietary system for cars something that, if they all pulled together, it could be for the benefit of all concerned. If people wanted to bring their entertainment into their cars via their smartphones, why not turn cars into digital environments altogether?

# Connected Vehicles

In 2009, the rapid convergence of car interiors becoming interactive spaces drove three separate industries – car manufacturers, a chip maker, and a mobile phone giant – into re-imagining cars as connected moving devices. I had started to work at Nokia innovation labs when my second start-up, Visual Radio, a radio signal and video streaming app cum CRM platform, was acquired by the Finnish telecommunications giant. At the time, Nokia Design and Innovation departments weren't even located at the iconic headquarters in Espoo, but in Ruoholahti, a Westerly Helsinki quartier created in 1910 by connecting several small islands with earth fill. Our department had a Star Wars-sounding name for a planet: NEBU, an acronym for New Emerging Business Unit. Our mission was to launch future products and services so radically removed from the traditional business of the firm that we were told to launch them under separate names and brands from the company's. We were

---

[1]Quoted from "Changing the In-Vehicle Infotainment Landscape" Copyright © 2010 GENIVI Alliance, All Rights Reserved.

23

looking for new commercial verticals for the firm, like wearables and sensor-based health-tech, so we created a partnership with NIKE and we launched their first version of a sensor-fitted running shoe and mobile app for activity tracking; when desktop widgets made their entrance into computer screens, we launched their mobile versions, "widsets," which later on became what we now know as mobile apps. The connected car got my attention, so I ventured into this road less travelled. Contacting car manufacturers on behalf of Nokia, which at the time was the world's most beloved handset maker, I was a welcomed party, and managed to understand many of the synergies that the automakers shared with us under one vision for the future: connected mobility.

A huge advantage for car manufacturers was the existence of a headunit in every car, a screen console that every driver was familiar with for at least a decade. It was a good pillar to build upon. By February 2010 during the Mobile World Congress in Barcelona, under the theme "Vision in Action," Intel and Nokia announced their partnership for a Linux-based operating system aimed at notebooks, tablets, connected TVs, and in-vehicle systems: *MeeGo*, a merge of their individual open source systems, *Moblin* and *Maemo*. It was Nokia's effort to find a vertical in the open source community to stay one step ahead of iOS (Apple) and Android (Google). Open source would guarantee seamless scalability to all new mobile devices, including cars. It would allow Nokia to have an alternative to their Symbian operating system, which still by end of 2010 was the OS of choice for smartphones worldwide.[2] When Microsoft's Stephen Elop walked through his office at Espoo two months later, on April 22, Nokia was the leading mobile phone maker for the world outside of the U.S., a market of 6.7 billion people. Elop arrived to oversee Nokia's acquisition

---

[2]"Symbian was used by many major mobile phone brands, like Samsung, Motorola, Sony Ericsson, and above all by Nokia. It was also prevalent in Japan by brands including Fujitsu, Sharp and Mitsubishi." Wikipedia https://en.wikipedia.org/wiki/Symbian

potential by Microsoft, and was quick to slim down the Finnish company of any unnecessary, redundant, and inoperable business. Symbian was the first to fall. By the end of February 2011, Nokia announced that it would switch to Microsoft Windows Mobile for all its smartphones while it wound down Symbian. Elop sent to all Nokia employees a controversial memo known as "the burning platforms memo" where the state of disarray and disparaging strategies had turned the leading mobile manufacturer into a sinking boat, or an "oil rig in flames" as he described. Elop did not only apply the brakes on Symbian. MeeGo was also abruptly halted with the same fatal stroke. Industry analysts, tech media outlets, and Intel themselves were kept in the dark about Nokia's sudden plans of ditching all platforms, and putting *MeeGo* back into the R&D projects activities. Somebody pushed a button and every single OS at Nokia was canned. Intel was left in the lurch and had to turn to the Linux Foundation for help. The OS was renamed *Tizen*, and the Linux Foundation pulled all the cards to get Samsung and other members involved in its development, including telecommunications companies Telefonica and Vodafone. The future was HMTL5[3] and everyone saw its potentiality: connected everywhere across every imaginable device, including TVs and cars. In the end, Samsung kept *Tizen* alive all the way up to the present moment. In 2015 the Tizen In-vehicle-infotainment (IVI) standard, which was for not just cars' but buses' and even airplanes' embedded computing systems, was migrated to Automotive Grade Linux. Rumor has it that Tizen still lives inside many of Samsung multimedia devices, but it is kept under wraps as to how regularly it is updated or even supported.

---

[3]HTML5 is a mark-up language to create mobile applications for all kinds of mobile operating systems, Android or iOS. HTML5 is browser based, and not a "native" app downloaded onto the cell phone, tablet, or multimedia device. Over time it became as fluid and seamless as native apps and thus the choice for many mobile developers.

# Mobility

By 2010, the ability of cars to connect to the Internet, make phone calls and operate routes via satellite navigation systems produced a new type of drivers: the rise of the *super-commuters*, workers who, on a daily basis, would drive hundreds of miles to commute to their offices, sometimes crossing over state lines or even entering neighboring foreign countries.[4] This new consumer presented different variations of motifs, mindsets, and attitudes for such emerging behaviors. In the United States, the 2008 financial crisis inflicted massive layoffs of large workforces. People with skills and career degrees also had mortgages to pay and were forced to search for jobs *anywhere*, not just locally or state-wide. The scarcity of opportunities forced people to sacrifice for their families: no one wanted to uproot children from schools, or worse, selling homes at negative equity, which was not an option most families wanted to even consider. Keeping a roof over one's family's head required self-sacrifice and guts to go out there and apply for jobs even if these were located hundreds of miles away. Commuting became almost a daily migratory journey for those who dared. Fortunately, at this point in history, cars were not only comfortable enough to drive for hours, but offered hotdesking facilities for those having to juggle conference calls or use the time to dictate memos to their staff, set self-reminders of to-do lists, or learn a foreign language. The hours spent driving were being leveraged to continue holding office hours on the move. In 2012 I wrote an article on how cars were becoming the next connected platform[5]

---

[4]The extent of cross-border commutes within European countries is such that the European Union has put in place a detailed system of healthcare, unemployment, family benefits, and retirement rights to allow citizens of one nation to work and receive such rights in a neighboring one. Commuting in the twenty-first century has expanded across borders as a form of daily immigration thanks to transport and the freedom to drive over the border with ease.

[5]Martinez, Inmaculada. "Cars; The Next Mobile Platform," *The Kernel*, May 25, 2012

on the occasion of Google getting approval from the California Senate for their autonomous car development and testing on public roads. The bill to make California the first state to allow autonomous cars trials was passed by a bipartisan vote of 37 to nil. Everyone voted in favor. *The Los Angeles Times* interviewed California State Senator Alex Padilla (D-Pacoima) for commentary. The senator highlighted the mantra that Google and other car manufacturers of autonomous vehicles were preaching to authorities and lawmakers: that autonomous vehicles would make roads safer, and even reduce traffic congestion. I saw an additional opportunity: people would use their time differently, leveraging from the spatial comforts of well-designed cars. Furthermore, people would stop being drivers and become passengers, engaged in other activities. On the socioeconomic level, the state of California had a keen interest in autonomous cars development for the simple fact that the state had not invested in railroads or public transportation as it should have. Being the typical European, at various points in my history of Silicon Valley visits, I would attempt to take the CalTrain to Palo Alto from downtown San Francisco only to discover that trying to return in the afternoon would be almost impossible, as the trains ran at odd and infrequent times. How absurd and costly to rent a car for just a day down at Sunnyvale for venture capital meetings. What a pain to pick up and return the car at the airport, so that it would be cheaper than at downtown locations, pay for petrol and parking. This was before Uber, as you have already guessed. Public infrastructure, like roads, bridges, and motorways, is a sign of economic health and social awareness: traffic is not only a health hazard for civilians – stress for the drivers and fumes for everyone living near a big highway. Still, many economies and geographies have relied on cars and the commuting for work as an assumed way of life that has evolved toward new user experiences, rather than dead-ends that will demand their demise. While it is true that cars are not the smart solution where it comes to transporting people and goods on a daily basis, on the same route, for a disproportionately large distance, they are the answer to allowing people to reach destinations that are remote and not serviced at the last leg, that is, the actual building, science park, or office complex. For workplaces outside of city centers, cars are the only way to make it sane and less haphazardly.

# Seamless Control

The user experience in the car commuting journey requires a dashboard capable of connecting to the Internet, navigating with satellite support, and being able to establish safe telecommunications without taking one's hands off the wheel. The first car that came to market with voice control was the 2005 Honda Acura. The Acura was a model that Honda began to manufacture in the mid-1990s and ten years on, the Japanese company was solidifying its market share toward the luxury sedan. Voice control seemed to them a distinctive differentiation. The Acura sound system was now MP3/WMA compatible and offered extras such as Dolby Pro Logic II decoding and speed-sensitive volume compensation, features oriented toward attracting music enthusiasts who were prepared to pay for this in-car sound experience. Honda partnered with IBM aiming to attract innovation credibility. For a first version of an in-car voice command system, the level of sophistication was considerable: the driver could control the climate temperature, the entertainment system, the navigation, as well as have access to other bundled services such as weather reports. When the car was reviewed by industry experts, the voice commands feature, something that no one at the time saw the point of, showed its limitations: the microphone was installed in the car ceiling and underperformed when background noises or even the air vents were at full blast. If drivers did not enunciate words correctly, the voice control system would just get into a tizzy. Early voice commands were still very far from being able to process free speech, so either the driver learned a predefined list of voice activation commands or the system would not work at all. The Acura's voice recognition went undistinguished and did not create adoption or the furore expected. From an artificial intelligence perspective, and having worked on chat bot voice command projects, I have come to understand why humans find it difficult to interact with predetermined voice commands: human dialogue is mainly abstract, free flowing, and above all, mostly not really about the words, but about the communication itself. When we speak, we tend to do it to establish a "connection" with another person, not just to read a list of to-dos. We do not "talk" to Alexa, as in, "having conversations with it." We force ourselves to speak to it in a manner that we do not speak to anyone else. Think about it. Many parents have realized that their toddlers were not learning basic communication skills because they started to talk back at them like they were taught to talk to home voice command gadgets. In anthropology, when we study how humans learn to speak we denote that speech is a form of integrating ourselves into our family, into our group, into society. Which explains why everyone working today toward machine voice communication is trying to embed this abstraction into the machine's learning environment. It is hard. Speech is more biological than cognitive or quantitative. Two years later, and this time in association with Microsoft, who was developing an auto operating system, Ford launched their own version of in-car voice commands: *Ford Synch*. The launch at the annual North American International Auto Show in January 2007 was flamboyant and pulled all the stops, involving the helms of the two companies, Alan Mulally, Ford's president and CEO and Microsoft's very own Bill Gates, the leader of enterprise software. The exclusivity period to Microsoft Auto OS was to run until November 2008, a period of time by which Ford aimed to develop its own proprietary technologies as well as involve other third parties in co-development. In later versions of Synch, a female-voiced text-to-voice system called "Samantha" would also be able to read text messages received while driving, a technology that most telecommunication companies were aiming to nail in order to increase their revenues for messaging services. This feature provided the driver with an incredible sense of command while at the wheel and became the precursor of a long-time coming dream in robotics: to make machines speak to us. Still, this service was dependent on the mobile network of choice and language compatibility was something that Ford could not resolve in bi-lingual countries such as Canada. The system could only operate in one language at a time and many English language commands did not exist in other languages. Technology's *lingua franca* is English and this did not lend itself to real natural language environments. In 2013 Skoda Octavia – *you read this well* – came to market, still not showing any major advances to what Ford Synch was offering. Six years on, no one was investing in this modality. Car manufacturers were at odds as to how they could recoup their investment in this type of technology that was very hard to deliver with the quality and expectations that humans demanded. No other gadgets in the consumer market seemed to have nailed this new feature – Amazon Alexa was successfully launched a year after in 2014, so no one had preconceived expectations or a comparable benchmark to appreciate if an in-car voice command system was better or worse than a home voice-activated electronic.

The aim to teach machines how to converse dates back to 1964 and the birth of "ELIZA," a chat bot created at MIT to showcase the "frivolity" of conversations that humans could have with machines who in the 1960s possessed the brain of a learning toddler and were obviously not very clever or amusing to have conversations with. Joseph Weizenbaum, its creator and one of the fathers of modern artificial intelligence development, built ELIZA as a natural language processing program capable of attempting the Turing test, or how a machine can trick humans into thinking that they are chatting to another human, not a machine. ELIZA was able to engage in scripted conversations but not to converse with true understanding because ELIZA was built as a retrieval system, that is, a system that generates predefined responses that heuristically appear to be appropriate based in the semiotics of the dialogues and their contexts. SIRI, still at times, is still a retrieval system and I have tested it to death regularly by asking: "Hey Siri, does God exist?" and noting down every one of its politically correct answers. I just did it again. Siri's answer: "It's all a mystery to me." Retrieval systems are carefully censored by the developers that build it. Ford Synch's "Samantha" was designed to interpret about one hundred shorthand text messages such as "LOL" and even read "swear words," but someone within the marketing team considered that this was about as far as the company would go to demonstrate its ability to connect with people and mainstream culture. For all other potentially obscene enunciations, "Samantha" was censored. Although rudimentary, retrieval-based methods offer certain advantages: sentences are grammatically correct, and there are no misspellings.

But this limits their ability to encompass human interactions because humans are fundamentally individual in their speech and unpredictable whereas retrieval systems tend to get stuck in conversations for which no appropriate predefined responses have been programmed. For this very reason, they are also unable to "remember" if something was mentioned earlier in the conversation. The real aim for AI natural language systems is to master the use of generative models, that is, teaching a machine to respond with its own choice of words and sentences to a given dialogue cue. This requires deep learning approaches to its training which in turn implies having vast amounts of data for such tasks and the use of appropriate reward functions for its "deep neural brain" to ascertain the concept and the context of what you are trying to teach it. It also requires time, computing resources, and the risk of the system of outsmarting you and your team by inventing their own grammar and version of the English language, which is what happened to Facebook when it attempted to teach two bots how to barter with each other. Very long, winded sentences tend to be agrammatical or incomprehensible sometimes because the longer the sentence, the more difficult it is to automate it. By the same token, certain short sentences or even monosyllabic responses by humans, such as "like," "I see," or any other verbal support to keep the conversation going, is still hard to teach a machine to interpret with full understanding, especially sarcasm, insecurity, or inability to get to the point. In the early days of voice-activated in-car controls, the dialogues between driver and system were close-ended: the controls were a handful, and the features a certain number. No one was expecting to have "free-flowing," open-ended dialogues with the car system or interactions that would derive into domains outside of the controls menu.

Human communication is hard to replicate with accuracy because it goes beyond words. When we speak, we express ourselves in very different narratives than when we write. Speaking to someone engages our bodies, the tone that we use, how we pronounce words, how fast or slow we speak, how we make use of verbal crutches such as "like," "you know?", and other endings that are simply intercalated in the conversation as checkpoints to ensure that the listener is empathetic to us, that their attention is still there. In conversation, and most importantly, in public speaking, the spaces and silences carry a weight and a power of connection like no written text could achieve. Speaking is profoundly human, extremely biological, and unconscious. It is not about words but mostly about everything else. In comparison, written communications, exempt of all biological input, rely heavily on words, on the perfect architecture of grammar. Written communications is like weaving or knitting a textile. It is a recipe for the perfect loaf of bread where a tiny mistake can leave your sourdough flat and unsavory. Writing is about painting a picture in someone's mind. Speaking is about befriending someone or proving that "we don't mean to kill them," as anthropologists say of non-verbal communications that we pepper our discourse with. Will we ever teach machines how to converse with us, how to master sarcasm and humor?

The automotive industry abandoned the voice recognition features for another four years, disheartened by the negative reviews and the technical impossibilities of achieving the kind of user experience that their customers were used to with everything else that cars had to offer. But the dream of commanding machines with our voice remains embedded in those of us who watched TV as kids and were mesmerized when captain Picard, commander of the Star Trek Enterprise spaceship, would ask the computer to materialize a cup of Earl Grey out of thin air or when 1980s TV sensation KITT – Knight Industries Two Thousand, a car that inspired a generation of *Knight Rider* fans to continue remodeling 1982 Pontiac Trans Am cars – shocks actor David Hasselhoff by asking him what kind of music he would like to listen to or if he would rather hear some information about Silicon Valley. Ironically, their very first mission together took them to the home base of both Apple and Google, car industry outsiders that have managed to create in-roads in the voice commands vertical.

Still, somebody, somewhere deep inside the product innovation team at Honda, had a light burning for the off chance of one day getting a car to do this. Robotics is a core industry in Japan and this could not go unchallenged. The failure of the Acura voice control system did not deter the team behind it and a year later an improved version was embedded in the 2006 Lexus system. Still, and in spite of newer functionalities added over the years, by 2017 the Gen 3 Lexus RX350 continued to have a traditional structured menu approach that the driver had to read, which defeated the purpose of using voice commands, since the driver was forced to take their eyes of the road in order to read an on-screen menu. A study conducted by the National Highway Traffic Safety Administration at the U.S. Department of Transportation in October 2016 entitled "In-Vehicle Voice Control Interface Performance Evaluation – Final Report" revealed that 25% of voice tasks were abandoned by drivers out of frustration. Since most car models after 2012 came equipped with some form of voice command system (VCS), the researchers were able to test how drivers reacted across a wide variety of car models and aimed to accomplish a suite of three typical tasks: a radio interaction, a navigation request and calendaring an event. The more complex the task was, the longer it took the driver's attention from the road. *"In the case of a navigation destination entry on a production vehicle, the supposedly hands-free and eyes-free operation led to an average task completion time of over 90 seconds and an average of over 30 seconds of off-road glance time"*[6]. This was disappointing. The industry needed to develop "pure" voice-based systems that would not demand a driver to be distracted with reading a screen. Car manufacturers like BMW were already developing a natural language approach to VCS but these systems, built on Cloud technology, suffered delays in the over-the-air (OTA) transmission, forcing drivers to check the screens wondering if they had made a mistake.

The report had a very accurate account of which human factors challenges prevented the VCS to deliver a good user experience. The type of tasks that we aim to achieve via a VCS are complex and require the interactions of various components: firstly, correct enunciation of destination in a navigation system, and audible commands within a vehicle that is moving at speed amidst traffic and other sound disturbances which may interfere with the correct speech recognition of the system. Secondly, its abilities to correctly interpret pauses in the discourse as part of the natural interactions between driver and VCS on the occasion that the driver would get distracted in mid-sentence or speak slower than average or drivers beginning to talk before the VCS is ready to analyze the speech. Lastly, the researchers noticed the failure of the system to adapt

---

[6]The Effects of a Production Level "Voice-Command" Interface on Driver Behavior: Summary Findings on Reported Workload, Physiology, Visual Attention, and Driving Performance", Bryan Reimer and Bruce Mehler, November 18, 2013.

https://www.researchgate.net/profile/Bryan-Reimer/
publication/258757659_The_Effects_of_a_Production_Level_Voice-
Command_Interface_on_Driver_Behavior_Summary_Findings_on_Reported_
Workload_Physiology_Visual_Attention_and_Driving_Performance/
links/00b7d528e555168c18000000/The-Effects-of-a-Production-Level-
Voice-Command-Interface-on-Driver-Behavior-Summary-Findings-
on-Reported-Workload-Physiology-Visual-Attention-and-Driving-
Performance.pdf

itself to individual driver differences, as not all drivers are technologically savvy and age can play an enormous effect on how adaptive a VCS needs to be in the future to guarantee inclusivity and diversity protocols. The report concluded that more optimization was needed in order to offer the safety that all drivers and passengers deserve.

# Onward

The years that followed saw the rise of voice assistants across mobile phones, and the strategic partnerships and alliances that Google and Apple began to establish with the automotive industry. As early as 2010 Apple started a collaboration with BMW Group's Technology Office USA, an alliance that was announced during their developer conference WWDC in June and where a hidden app within their newly minted IOS 4 called "iPod Out" hinted at their plans to develop a car-specific interface for apps. Called "PlugIn," it first shipped in the 2011 BMWs and Mini series cars, evolving from the inadequacy of previous interoperability standards such as MirrorLink that were developed to integrate smartphones and car infotainment systems. If MirrorLink or a car manufacturer's own system was going to pull your music library from your sacred iPhone, the Apple monks were horrified with how awful that data ended up when displayed on the car head unit. Apple being Apple – design-obsessed, proprietary-paranoid – took up the challenge of designing Apple-compatible car systems. Their move did not go unnoticed and Google went for the same challenge: a market share of the car infotainment business of the 2010 decade.

The year 2014 can be considered the starting point of voice commands innovation. It was the year that Amazon gave the world *Echo*, a smart speaker system connected to *Alexa*, the first home voice-activated assistant that came to delight a consumer society mad about *Harry Potter* and the notion of having things magically happen. Not only did it handle the playing of music tracks, or read the daily weather prediction, but also ordered you an Uber, and built to-do lists, grocery shopping lists, or anything that could instantaneously be ordered on Amazon. Finally, a voice assistant that could personalize the shopping experience to an almost frictionless context: just say it, and a brown cardboard box would deliver it twenty-four hours later. Google, who traditionally has been fond of having all ten fingers in as many pies as possible, started the year announcing at CES the Open Automotive Alliance, a collaborative partnership between car manufacturers and Google to adopt Android within the connected car in-vehicle entertainment systems. In the background, what Google was developing was *Android Auto*,[7] a mobile app specifically built to interact with a car's head unit and controls systems. But this was just an opening move on the chess board. Within a week of CES, Google announced the acquisition of *Nest*, an Internet-of-Things innovator of smoke alarms and home thermostats, for $3.2 billion cash. What Google saw in *Nest* was the hardware that could deliver their software emporium, and dominate in an emerging new field within the consumer market: artificial intelligence. Literally, a fortnight after the *Nest*'s transaction, Google was acquiring an unknown British startup called *DeepMind*, whose founders had apparently developed and trained the most sophisticated artificial neural networks via deep learning programming. Google, who a month earlier had also acquired robotics company *Boston Robotics*, was now in possession of one of the most competent AI teams in the world – who soon joined forces with the existing machine learning and language processing human capital in Mountain View, and the two hardware verticals to disrupt the future: robots and home IoT devices that would sentiently gather consumer data 24/7. Check-mate. Five months later in June, Android Auto debuted at Google I/O, their developer conference in

---

[7]Volvo's first EV will run native Android. The electric XC40 SUV will be unveiled next week www.theverge.com/2019/10/9/20906777/volvo-xc40-suv-ev-native-android-auto-google-assistant-maps

Mountain View, California. Where was Apple when all this was occurring? Although it started the race with a handful of years heads-up and a heritage luxury partner in IBM, the Apple CarPlay project became the victim of Apple's hardware-paranoia and suffered release dates delays. Still, Apple launched CarPlay at none other than the Geneva Motor Show in March with Ferrari, Mercedes-Benz, and Volvo. In true Apple fashion, the slap in the face to Google could not have been more obvious: Google, who in 2007 had come up with the Open Handset Alliance to attract handset manufacturers to its mobile platform, was replicating the same formula with the Open Automotive Alliance, in the hopes of ensuring a place in the automotive industry for its Android Auto. Apple, elitist as per usual, was going for its typical market share tactics at the peak of the consumer pyramid, where the luxury companies look down on the rest of the retail mortals. A year later, Hyundai became the first car manufacturer to offer Android Auto support[8] while doting their luxury models like the Sonata Sedan with CarPlay. The car industry had that purchasing power: cars, at least in the 2010s, were still consumer giants who could beat the Silicon Valley private tug of wars to dust by picking and choosing whomever they fancied working with. Also, a handful of automakers had developed their own proprietary systems[9] or smartphone syncing[10] in parallel to whatever the mobile industry was aiming to sell to them.

---

[8]According to Android Auto website the list of manufacturers supporting this VCS platform is long and specifies for which models at every automaker: Abarth, Acura, Alfa Romeo, Audi, Buick, Cadillac, Chevrolet, Chrysler, Dodge, Ferrari, Fiat, Ford, GMC, Genesis, Holden, Honda, Hyundai, Infiniti, Jaguar Land Rover, Jeep, Kia, Lamborghini, Lexus, Lincoln, Mahindra and Mahindra, Maserati, Mazda, Mercedes-Benz, Mitsubishi, Nissan, Opel, Peugeot, RAM, Renault, SEAT, Škoda, SsangYong, Subaru, Suzuki, Tata Motors Cars, Toyota, Volkswagen, and Volvo.

[9]Lada, Tesla Motors, and Honda (only their Gold Wing and Africa Twin cars offer CarPlay).

[10]BMW ConnectedDrive, Hyundai Blue Link, iLane, MyFord Touch, Ford SYNC, OnStar, and Toyota Entune.

The evolution of these two mobile operating systems thus took a usual pattern: Google would go for horizontal expansion while Apple focused on achieving "industry firsts"[11] with pedigree automakers, both offering their apps in mobile devices that would connect to a car head unit where the exclusive partnerships would not allow for either system to come direct from factory at a given automaker. The success of Alexa inflated Amazon's bravado in approaching certain car manufacturers like Spain's SEAT and VW Group to sign deals for in-car systems speech recognition. This was an anomaly in an industry that had spent a decade dealing with Google and Apple, but the preliminary pathway to a trend that will see its rapid mainstream expansion in the 2020 decade: the extension of the home into the car cabin. My personal opinion is that the long-distance runner that Amazon is has met its match in Google. The Lord Retailer that Amazon is has a predetermined field of influence: consumer purchases. Google, on the other hand, has a different aim long-term: to know absolutely everything there is to know or assume on every single human being on the surface of Earth, to profile and segment each individual and to serve each of us ads and ad-to-purchase opportunities that can be derived by Google's intimate knowledge about individual subjects. While Amazon made in-roads into the voice command automotive vertical and Apple was eating cake with the car royalties, Google kept on pushing Android Auto into optimized versions of itself and with APIs that would comprehend a wider range of sensor-based activities within the vehicle. Leveraging from its years of developing Maps and acquiring the people's choice traffic app Waze, mobile consumers, whether they were Android or iOS users, would probably choose Android Auto in their cars to enhance their navigation routes and overall in-vehicle entertainment experience.

---

[11]In 2017 Apple went the way of "trucks," with Scania's new models becoming the first heavy duty trucks in Europe to support CarPlay and Volvo VNL in the United States.

In parallel to this sidelines action, on January 3, 2018 at CES in Las Vegas, a "usual suspect" in the auto industry disruption arena launched their in-car voice assistant, ticking off all the boxes in the wish-list that the U.S. Department of Transportation had put together in their 2016 report and more. Dr. Dirk Hoheisel, Member of the Board of Management of Robert Bosch GmbH, announced without hesitation that Bosch was "putting an end to the button chaos in the cockpit." and that the company was turning "the voice assistant into a passenger." Bosch's VCS was a natural multilingual system that did not require an external data connection for support, greeting the driver with a friendly "I'm Casey (or "Linda," or "Michael," as it is the driver who decides on the name for the Bosch voice assistant), your new passenger. Are you ready to get started?" And were we ready! At long last, a car would acknowledge our presence and make itself useful from the get-go. Germany had had its own VCS audit, conducted by Allianz Center for Technology, and the results of their report were a mirror image of the one conducted in the United States: drivers were distracted when operating navigation systems, adjusting the temperature of the car, or answering telephone calls while attempting to do this via a VCS. This was extremely unsatisfactory for an industry that had spent years making cars safe to travel in.

Bosch's VCS allowed drivers to say whatever they needed in whichever way they wanted to say it. No more memorized commands. It had learned to recognize natural language, which is how humans speak. You could say, for example, "Casey I'd like to send a text to Jenny" instead of giving it an order in the usual robotic style "Casey text Jenny," which you could also say, but it made you drop your good manners. The system recognized accents and dialects as well as up to 30 foreign languages used interchangeably. Say, for instance, that you were driving in Baja California, where the street names are in Spanish. No problemo. You could speak in English and ask for directions to any locality whose name is in Spanish. Casey had been trained over a decade to distinguish the nuances of languages. It could even assume contextual scenarios.

If you wanted Casey to call "Paul" and it was a Monday morning and you were not too distant from your office, and Paul was a colleague or a client that you usually called in those circumstances and geolocation, Casey would call "that" particular Paul on your contacts list. This was as "earth-shattering" as when Amazon launched in 1998 its now famous "collaborative filtering" algorithm for personalizing your book shopping experience. I know the name of this "algo" because I also built a similar one two years later for the early days of the mobile Internet. It is an object-oriented database that learns to make sense of the choices a human makes of a given system based on time, location, and previous behavior. It is what we ended up calling a "recommendation engine," suggestive personalization or a simple AI system that advances your needs. "People who bought this book also bought these other ones" is now a thing of the digital past, but it is what opened the Pandora box to customer-centric services and the notion that humans need to be dealt with deference, attention, and the extra mile. This is what distinguished Amazon from the rest of the book vendors and what has turned the company into an outlier. In addition, the off-line feature really provided a new concept for edge computing and data communications. How many times have we entered a tunnel or driven through a black spot in the cellular network grid losing connectivity? As described earlier, VCS are cloud-based systems that send and receive data packages over cellular networks and if these are fragile, the whole operation goes out the window, frustrating both drivers and car manufacturers. To resolve this challenge that laid outside of the automotive industry's remit, Bosch built a storage capacity for all processed commands which remained in the system until the connection was reestablished. In edge computing we can perform AI inference activities within a small computer off-line and with real live data because the algorithm has been previously trained in the Cloud, in the super capacity servers and with terabytes of data. Was this VCS a demonstration of sorts? Because it is hard to find any current information as to which vehicles currently run the Bosch VCS, something that puzzles me, given the amount of work and R&D budgets that it may have cost to develop it, but it may well be that this cabin feature has not gained any momentum within a current industry prioritizing the reduction of their fleet's $CO_2$ emissions rather than seeking to enthrall customers with a voice command box of wonders.

Beyond the voice recognition gap of opportunity, both mobile and tech companies have continued to make in-roads into verticals where they could add value to the user's car experience. One of those coveted realms is the Car Connectivity Consortium, a cross-industry organization where automakers, OEMs, and other technology suppliers to the mobile industry work toward the creation of smartphone-to-car standards for two long-standing projects: Digital Key – where drivers will store their digital codes to their cars and engine ignition in their phones – and Car Data, a platform to gather vehicle usage data which may be leveraged for reducing insurance premiums for good driving, road monitoring, and fleet management. This was the organization that launched MirrorLink, the most adopted and deployed open standard for connecting apps between the smartphone and the car. Intriguingly, though Apple sits on the board, it refrains from showing in the general list of members. It is the paranoia that we have grown to love and to hate at times. Still, Apple has a bee in its bonnet and it is sentiently working toward its own approach to make the driving experience become an Apple family user story. In June 2020, at their annual developer conference WWDC, Apple announced their launch of a digital car key that iPhone and iWatch users could deploy to unlock and start their cars.[12] Not Earth-shattering and not an innovation per se, because the car industry has been using Near Field Communications (NFC) radio frequency to do the same since the 1980s, and the new emerging frequency, Ultra Wideband (UWB) has been around for decades, but a fundamental development that proves that vehicles, in this decade,

---

[12]Continuing with their partnership with BMW, the first car to display this feature will be the new 2021 BMW 5 Series.

are becoming increasingly closer to what consumers consider a rounded experience, a seamless flow of technological convenience where all their gadgets converge into a single narrative. If there was ever a company hyper-focused on the all-in-one, frictionless tech experience, that will be Apple. Just like mp3 turned vinyl records into digital files, Apple elevated this to the iPod experience, and then crossed it over to make iPhone the luxury lifestyle gadget for those who sought the all-encompassing hardware/software experience. Apple installing UWB chips in their iPhone 11s and iPhone 11 Pros in 2020 is not their cheeky way of winking at the car industry with a car fob app, but a signal that a deviation toward wider convergence between the driving and the mobile experience is beginning to make sense to them when they work on the life cycle of their own products. For example, you could store all the digital keys of all the cars in your family on your iPhone. Imagine the joy of never having to look for keys around your house, handbags, or fear that you absentmindedly dropped them in the street. If mobile phones become the de facto platform where we run all the gadgets of our lives, this centralized prime position will strengthen the foothold that mobile handset manufacturers exert on consumers. Apple car, you said? This may take a while longer to see in the market. The apparent Apple and Hyundai-Kia talks to produce a fully electric, fully autonomous vehicle for the Cupertino company out of the Kia assembly plant in West Point, Georgia, have died out. It seems that some senior executives at Hyundai did not fancy themselves becoming another Asian manufacturer for the intransigent Apple, which at times has driven its Chinese manufacturing provider off the rails with insane deadlines, creating bad press with news of Foxxconn employees committing suicide and hundreds of them succumbing to work exhaustion. On another level, can the Apple brand branch out to vehicles? Even Google gave up. I personally do not see consumers buying an Apple car. For what, to listen to iTunes or talk to the useless Siri, who cannot even deliver a proper conversation? Does that mean that Samsung is also thinking about entering the automotive industry? Apple tried to make us all believe that it could become the next TV until we all bought iTV and quickly came to the realization that we had been conned into buying a glorified device that delivered what our iPhones or laptops could do. Apple TV+? I will not be renewing my annual subscription to a studio that has delivered a mere handful of inhouse productions and 99% of the time is a glorified iTunes with titles retailed at prices higher than Amazon Prime. Do not believe the hype: Apple will not go into autonomous vehicles, or electric ones. It will be lucky if by 2025 its market share continues to be the same.

# Summary

Turning cars into contextual spaces of human activities – from listening to music to conducting business or managing car features via voice commands – kick-started the digital revolution that elevated vehicles beyond their transportation duties. It was inevitable that vehicle cabins would evolve to host other human activities as the mere act of moving from A to B would be safer, easier to handle. What was unexpected was the rate of transformation. Digitization has that power of exponential disruption that catches some unaware, and others ready and up for the challenges. Before we even considered self-driving cars, it was us the ones that started to be uninterested in the driving itself, the ones that wished we could turn vehicles into spaces of multiplicity as technology and digitization tempted us with contexts of convenience and delight.

# CHAPTER 3

# The 5G Car

*With the upgrade of telecommunication technologies, vehicles are bound to evolve toward connectivity scenarios that will enhance our lives, not just in terms of traffic safety but also how cities are re-imagined into urban centers that will improve the lives of citizens, our daily activities, and the flow of life, work, and play.*

## The Mobile Revolution

Cars were fitted with telephones ever since 1956. These could handle calls without the need for an operator to direct the call but were not truly "mobile." They were dependent on a specific telephone mast that acted as a base-station which was only able to handle communications within a very reduced geofence.[1] In order to offer communications not bound by geography, the first generation of mobile networks was an analogue technology developed in the 1980s. Various technologies were deployed according to each country or region: NMT (Nordic Mobile Telephone) for the Scandinavian European countries, Switzerland, the Netherlands, Eastern Europe and Russia; AMPS (Advanced Mobile Phone System) developed by Bell Labs and Motorola for North America and Australia; TACS (Total Access Communication System) in the United Kingdom; A-Netz to E-Netz in West Germany, Portugal and South Africa; Radiocom 2000 in France;

---

[1]In mobile telecommunications a geofence is a virtual geographical perimeter.

© Inma Martínez 2021
I. Martínez, *The Future of the Automotive Industry*,
https://doi.org/10.1007/978-1-4842-7026-4_3

TMA in Spain; and RTMI (Radio Telefono Mobile Integrato) in Italy. In Japan, a country that led the early mobile innovations, each of the telecommunication companies launched their own standards. JTACS (Japan Total Access Communications System) by Daini Denden Planning, Inc. (DDI) and the TZ-80n series by NTT (Nippon Telegraph and Telephone Corporation).[2] What these 1G networks were able to achieve was one of the most important pillars of mobile communications: *roaming*, that is, the ability to place calls to a mobile telephone within a network, irrespective of its geographical location. The second generation brought to market the first digital systems, deployed in the 1990s introducing voice, SMS, and data services. The primary 2G technologies were GSM/GPRS & EDGE, CDMAOne, PDC, iDEN, IS-136 or D_AMPS. Unfortunately, the telecom operators at the time did not see much business in the data side of the mobile industry. It took almost two years for Vodafone in the United Kingdom to create a pricing platform for their text SMS services,[3] for example. Data was still a feature of a mobile network, but not a major earner. In 1994 GPRS (General Packed Radio Service) began to be developed into a standard by the Special Mobile Group (SMG) within the European Telecommunications Standards Institute (ETSI). GPRS was coming to GSM infrastructure to provide reusable end-to-end packet-switched services, thus not demanding the networks to be upgraded. Three years later, GPRS specifications were approved and by 1999 it was a service that was ready to be commercialized. GPRS could run data downstream speeds of between 56 to 114 kilobytes per second. At the time, this speed was only good enough for basic WAP portals, like *Yahoo! News* and to download your emails to your phone, but not to deliver image-heavy websites, let alone what some mobile operators

---

[2]Source: 3rd Generation Partnership Project (3GPP) website

[3]To the delight of their customers, who massively used the service free of charge. People stopped calling friends and family and sent them SMS texts instead.

claimed as "surfing the web on your mobile phone," a marketing claim that was very far from the true user experience. Although throughout the early 2000s many WAP developers threw themselves to the task of creating incredible gaming and e-commerce WAP sites, the whole ecosystem collapsed because of a series of unfortunate circumstances: WAP made the wireless transport layer security vulnerable, and claimed to support all wireless networks, even those of Motorola's FLEX and ReFLEX networks, which were created to run pagers and text messages communications, but had never been designed for Internet protocol applications. WAP was over-hyped and the cellular networks did not see how they could recover their investment in upgrading the networks to faster speeds of 2.5G or 3G. Then, by hook or by crook, and very much by a certain obsessive dreamer of tech called Steve Jobs, a handset called iPhone ignited the hearts of every single consumer in North American and Western Europe in 2007. By then, the networks were stable and run 3G data speeds good enough to deliver the smartphones' promise of streaming music, Internet anywhere, and push emails. And with Bluetooth, a radio frequency for nearby objects, the Internet was finally mobile. The issue then was to handle the increasingly large amount of cell phones and other devices connected to it concurrently, downloading stuff at 2Mbps (megabytes per second). On December 14, 2009, Scandinavian network operator Telia-Sonera launched the first commercial 4G network in the capitals of Sweden and Norway – Stockholm and Oslo – and a year later other European countries began to upgrade their 3G networks to either 4G or what was called 4G Long-Term-Evolution (LTE), a bit slower than true 4G. The difference between 3G and 4G networks was radical. While speeds in 3G networks were achieving kilobytes per second, 4G brought with it something supersonic: megabytes of data speed, achieving downstreams of up to 5Mbps, a speed comparable to what one gets at home via a cable modem or Digital Subscriber Line (DSL) of a telephone network.[4]

Mobile life skyrocketed to infinity and beyond with 4G and LTE. Still, there are great differences between each other where it comes to the standards assigned to each by the International Telecommunications Union Radiocommunications Sector (ITU-R). Whereas the consumer was awarded 4G speeds spectrums of up to 1Gbit per second, for connections considered for pedestrian or stationary "low mobility" contexts, and vehicles were awarded just 1Mbytes/second for "high mobility" needs, the crude reality was that network operators never delivered these speeds in real terms, and LTE was just marginally faster than 3G in some countries. This is why with 5G, the ante stakes will go much higher and will set the mobile experience worlds apart from previous network generations.

Cellular networks became faster because telecommunications companies upgraded their signaling towers. Where a 3G tower could offer a stable and fast signal to about 60 to 100 cell users, a 4G tower could do the same but for 300 to 400 people concurrently. But here is where the line between them and 5G is drawn because 5G is just another crazy level of everything. Spectrum-wise, while both 3G and 4G can easily operate within the low-and-mid bandwidths that range between 600 Megahertz and 6 Gigahertz, 5G takes off and flies solo all the way up to its unique 100 Gigahertz capability. What does this mean in real terms? It means that it can handle every connected device, not just mobiles, tablets, automobiles, and the 2010s Internet of Things, but also the Internet of Everything: connected vehicles, connected homes, and all that there is in connected cities.

---

[4]Within the existing network infrastructure, telecommunications companies were able to offer high-speed Internet connectivity using the free frequencies not used for voice communications. DSL connections can be symmetrical, that is, offering equal amounts of data uploaded or downloaded, or asymmetrical – the most popular, with much bigger allowance for downstream data speeds than for upstream, which seems to be the convenience most people require of the Internet.

The digitally linked society of the future that many of us dreamed of finally found its Godspeed. The decade that we are just entering is one where telecommunications companies will still offer every generation of networks, from 1G to the newly minted 5G and beyond. This is because society will require different spectrums for different connectivity needs. Lower spectrum bandwidths like 2G and 3G are more reliable and their signal is powerful enough to penetrate buildings. This is why when your 4G signal fails in the elevator, your phone automatically switches to 3G. It is therefore the preferred signal for edge computing, the new universe of connected devices within a closely knit location. In this realm, the Internet of Sensors, artificial intelligence inferences are performed right on the edge, without uploading all the data to the Cloud. The way Cloud and Edge become environments for AI is based on when we train algorithms and when we put them to work with live data. We train algorithms in the Cloud, where enormous volumes of data help us ensure that machine intelligence does comprehend what we are teaching it, and the algorithms perform with zero-error, delivering non-bias results. Once these algorithms are confirmed and trained, they are embedded into Edge devices such as mobile phones, tablets, low computational software systems, and cars. In the Edge realm, the data that they handle is live data, and in much less quantities. This is why, being able to perform algorithmic functions within this dual scenario – local edge for fast, real-time applications, and uploaded to the Cloud for algorithm training – is how automobiles became extraordinarily intelligent machines.

# Intelligent Vehicles

5G cars are, in real terms, computers with wheels. They come to market with an intelligent system ready to make decisions in real time. How this works out is very simple: the car's AI system will have a baseline knowledge of what it is supposed to perform acquired via machine learning training – which takes an awful long time to get right. This trained model is embedded into the car's system and once it starts to interact with the real world, it will make decisions based on inference, that is, predicting events based on new data inputs, and how it was trained to think. This allows for decisions that are easy to turn around and is ideal for applications such as computer vision, voice recognition, and language processing tasks, the baselines of the car of the future. A great example of this is the current safety measures in high-end car models deployed to ensure a driver's correct handling of the vehicle. The car is fitted with a camera sensor that keeps track of the road markings and will detect if the driver is steering erratically – perhaps from tiredness, or worse, a drunken stupor. At this point the system will infer that the driver must be alerted and will make the steering wheel vibrate or sound an alarm, whereas it will do nothing if the lane separation marks are crossed over when overtaking another vehicle. How will it know the difference? The driver uses the blinkers before maneuvering, or any other action built into the training model to signify that drivers are in control of their actions or following normal protocols. A car's headlights go on when entering a dark car park, or when dusk turns into night; high beam lights are switched to normal nighttime lights if an oncoming vehicle appears in our horizon in order to avoid blinding the other driver. All of these automated behaviors have been trained at factory level and the car, a computer on the edge of the network, performs them in real-time within milliseconds.

Before cars are able to be fully autonomous, there is a step in between our current vehicles and what the automotive industry is planning to offer: it is called the "connected car," a vehicle that connects to other vehicles, devices, and infrastructure. The automotive industry has been working on this for decades and via a multitude of approaches. A portfolio of technologies has been tested, proposed, and implemented, presenting such an array of possibilities that manufacturers have had the opportunity to pick and choose to create co-development alliances with different technology providers. In reality, all of them constitute "moonshots" at solving the safety issues around driving and managing traffic. Some are more scalable and less costly than others, but all aim at potentially working out in parallel, as additional layers of innovation. In the last twenty years the boiling pot of connected car innovation has prompted some very interesting technologies worth considering, but nothing will compare to what vehicle connectivity communications will bring in terms of game-changing scenarios.

# The Internet of Vehicles

Dedicated Short-Range Communications (DSRC), a one-way and two-way wireless communication specifically designed for automotives, was developed in the late 1990s and meant to be used by *intelligent transportation systems (ITS)*. Initially, the main purpose of ITS was to become a preventative measure to decrease the number of traffic accidents which, according to the World Health Organization (WHO) causes about 1.2 million deaths and about 50 million of injured-for-life people a year worldwide. This data puts driving accidents in third place among all causes of mortality in 2020, a whopping increase from the 9th position that vehicle deaths used to cause in 1990, when we drove cars that were less fitted with safety measures. As car ownership became more accessible, so did the number of vehicles, which in just three decades has grown to account for 1 billion of passenger vehicles worldwide, according to the International Organization of Motor Vehicle Manufacturers (OICA). Still, safety is one of the oldest mantras in the automotive industry and what keeps its legality and positive relations with governments, so the pressure to decrease accidents is of utmost importance.

Vehicles fitted with DSRC would broadcast their location and identify themselves individually to a monitoring system. One can use them to electronically collect fees in toll roads, or manage the flow and schedules of public transport. DSRC can even synchronize the individual cruise controls of a fleet of vehicles via what is known as the Cooperative Adaptive Cruise Control (CACC), which realizes the distance between the vehicle in front of yours and synchronizes the cruise speed accordingly. These type of vehicular communication systems form computer networks in which vehicles and sensor-based nodes along the roads communicate with each other, providing safety warnings and traffic information. In addition to speed and direction, DSRC can also give vehicle localization by a centimeter-base. If an accident occurs, ambulances can be dispatched instantly to the exact location and traffic can be diverted to other routes before they pile-up the area, jamming the roads and preventing medical aid to reach people in need. Cars can send each other warnings as to when to brake unexpectedly before the driver has a visual aid to do so. DSRC seemed to have a splendorous track toward adoption. ABI Research was predicting in 2014 that, by 2018, 10% of worldwide shipped vehicles would be fitted with DSRC and a 70% adoption share by 2027, until the telecommunications companies asked to share the 5.9 GHz band for connecting services that they planned to launch. After initial power-wrestling in the boardrooms of both car manufacturers and telecom companies, the spectrums were agreed on the understanding that together they could gain more rather than fighting each other at regulator's offices, the path toward 5G services was cleared.

# 5G: The Internet Superhighway

Connectivity is a crucial milestone in vehicles' evolution toward digital transformation because it offers a wide range of possibilities. Even within current 4G networks, cars have connected to each other (Vehicle-2-Vehicle, V2V) in order to map out road traffic, and to infrastructure (Vehicle-2-Infrastructure, V2I) in emerging edge computing scenarios based on the Internet of Things. But with 5G, the landscape will upgrade to superior levels of automation. Concepts such as "automated parking" whereby, upon entry into a smart parking building, your car will be "taken over" by an automated valet system which will find an available spot for your vehicle, drive to it, and park it – all of this while you are sitting at the steering wheel comfortably checking your WhatsApp, returning emails, or applying lipstick. This will allow drivers to get accustomed to completely letting go of their car, thus breaking the ice toward complete trust in an independent machine at speed. The Vehicle-2-Infrastructure (V2I) landscape will open to relevant IoT services that will digitally shape localities into becoming intelligent towns (smart cities of automated self-management). Vehicles will calculate speeds according to traffic lights switching, send signals to pedestrians to announce their nearby presence and prevent accidents (Vehicle-2-Pedestrian) as well as connect to the network (V2N) in order to plan more efficient routes in real-time, establish less polluted itineraries, and route through safer and slower streets that the elderly drivers will handle better. In the 5G environment, the digital reality will be one of a *connected to everything vehicle*, or Cellular Vehicle-to-Everything (C-2VX). Connected not just to each other but also to everything around them, vehicles will navigate a virtual reality of 360 degrees' self-awareness, functioning within two and three-dimensional realities as well as predicting future eventualities, reacting before they happen. The road safety of the future will upgrade to a new dimension with 5G networks, paving the way to the future of autonomous driving.[5]

Creating the necessary infrastructure to support this vision has required enormous OEM programs among telecom operators and infrastructure providers. The 5GAA (5G Automotive Association) aggregates eight of the nine global automakers, nine of the top ten global telecommunications companies, as well as top automotive suppliers, smartphone manufacturers, semiconductor and wireless infrastructure

---

[5]At Ericsson's D-15 Labs in Silicon Valley, 5G latency in vehicles is being tested with Radio-Controlled (RC) hobby cars and 5G hobby cars connected to a 5G network. Which one do you reckon handles better? www.ericsson.com/en/blog/2020/12/5g-latency-test-rc-hobby

companies, and test and measurement companies and the pertinent certification entities. It is an ecosystem of ecosystems working in alliance to develop solutions for intelligent transportation, mobility services and smart cities in the 5G 2020 decade. It is also an integrated and coordinated approach to roll out autonomous driving, define and agree upon standards, test prototypes within well-defined scenarios, and get ready for initial deployments around the world. Founded as early as 2016, the 5GAA has adopted a clever approach to working with regional standards[6] to define applications on a global scale. Creating a proven know-how library of case studies, different tests and OEM demonstrations were first deployed in Europe in 2017, and in the United States and China in 2018, having gained incredible traction before Covid-19 took over the world in the spring of 2020. One of the most crucial objectives was to create interoperability among the automakers, so that any vehicle could benefit from new use cases developed across the regions and defining application layer-specific minimum requirements for new messages between cars and infrastructure. The longer term expansion of the roadmap envisions developing toward industrial IoT, enterprise and automotive networks, private networks, and even environments of unlicensed spectrum from 2023 onward. The most important element to always bear in mind is that we are creating telecommunications between objects that move at high speeds. This is a challenge that goes beyond the static nature of traditional telecommunications and a requirement that demands safety at all times as well as versatility of interactions (V2V, V2I, V2P), guaranteeing latency performance and deploying higher spectral efficiency at speed. Basically, a five-ring circus of real-time multi-signaling, with the added pressure of handling people's lives inside fast moving vehicles.

To put these advances in automotive and telecommunications into an economic perspective, the automotive V2X market is estimated to be worth around US$ 689 million in 2020 and projected to reach US$

---

[6]SAE for North America, ETSI ITS for Europe, and C-SAE/C-ITS for China

12,859 million by 2028, at a CAGR of 44.2%,[7] of which the Dedicated Short-Range Communications (DSRC) segment will be the one with the fastest and most explosive growth within the V2X market, currently led by major players such as Robert Bosch GMBH (Germany), Continental AG (Germany), Qualcomm Technologies, Inc. (United States), Autotalks Ltd. (Israel), and Delphi Technologies (United Kingdom). Governments and regulators have played their part in enabling the growth and attractiveness of this sector by allocating specific spectrums to DSRCs. While in Europe the European Telecommunications Standards Institute (ETSI) allocated 30 MHz of spectrum in the 5.9 GHz band, in the United States the Federal Communications Commission (FCC) doubled the allocation to 75 MHz of spectrum in the same GHz band, signaling how bullish the government is in capitalizing this new emerging technology, pushing the US vendors to be as competitive as possible over the European ones. According to a report by Industry Analytics Research Consulting[8] [quote] "the overall market of DSRC On-board units and roadside units in the year 2017 amounted to be $100.6m. Also, the global DSRC market for passenger vehicles was estimated to be $60.7m in the same year and is estimated to grow at a CAGR of 6.8% for the forecast period." [end quote]. According to this report, the determining powers of influence of this sector, in addition to government support and encouraging regulations, will be the increased awareness for safety measures that the deployment of DSRCs will bring to drivers, an augmented reality in which vehicles will "communicate" to other vehicles and infrastructures, even under extreme weather conditions. This short range of communications increases the safety features beyond what the human eye and capacity for driver's reactions were able to achieve in twentieth-century vehicles, making "Vehicle Awareness" the cornerstone of future autonomous driving and the

---

[7]Automotive V2X Market. Published Date: Feb 2020 | Report Code: AT 5947
[8]Report Code: ATR 0032

requirement that smart cities and municipalities want to see operating on their roads before they allow for fully automated, autonomous traffic.

The infrastructure for these short-range communications requires not just the on-board units (OBUs) of each car, something that the vehicle manufacturers have been working on for years, but roadside units (RSUs), which each municipality must install as part of the public infrastructure. Whereas in vehicle-to-vehicle communications RSUs were not needed, this part of the equation, the allocation of public funds or private contracts to ITC vendors to build RSU infrastructure will determine which cities take off in testing autonomous vehicles and which ones will lag behind. Moreover, the potentiality of certain urban centers to become "intelligent cities"[9] will leverage DSRC infrastructure to increase public safety and traffic management beyond what can be achieved today. Dangerous circumstances such as not spotting blind spots while driving will be solved thanks to DSRC technologies. Drivers will be warned for forward collisions caused by the sudden braking of drivers ahead, or unexpected vehicles crossing intersections. Vehicles digitized with DSRC will be inspected for safety with less room for error, detected by emergency vehicles on a map for exact location, and become, in the next five years, integral parts of the Edge computing context of the future of life, work, and play. Gartner has taken the Cassandra side of this argument and predicted a doom and gloom 2023 future in which "30% of smart city projects will be discontinued" for being too techie, too keen on displaying emerging technologies that serve very little real purposes and are mostly a show-off of technology vendors, as it has happened in city projects that forced municipalities to enter into smart

---

[9]This new emerging sector is being addressed by automotive firms and venture capital funds dedicated to re-imagining the future of cities where it comes to connectivity. If you wish to stay on top of the latest, follow Urban-X, the urbantech accelerator and its Demo Days and collaborations www.urban-x.com/

city deployments that were technologically achievable but failed to derive any citizen benefit. Welcome to a project that I was part of in the Brazilian city of Rio de Janeiro, just when the municipality was preparing to host the 2016 Summer Olympics. Two technology giants, and allow me to not mention any names, convinced the city police department to create a crowd monitoring system that would spot crime on the streets in real-time, allowing the authorities to prevent potential muggings, arrest perpetrators and other law enforcement activities that would benefit from video camera footage. The project was successful. There, in the mission control room at the Rio Operations Center (COR), a combined effort between the Center for Integrated Command and Control (CICC), and the Army to coordinate the security operations, a multitude of TV screens showed us strategic shots of city intersections, accesses to metro stations and the Olympic stadium. And right there, via live feed, we could see all kinds of law infractions, so many that the police department could not handle a fraction of them, to the frustration of the authorities that felt shamed by a technological platform for not having enough law enforcement officers on the streets. What Gartner is pointing out is that human centricity needs to be put on the agenda whenever a smart city is being envisioned, and that the involvement of municipalities in technology projects has to ensure the direct benefits to the city dwellers.

# Edge Computing Mobility

There is an additional layer of value in the 5G network proposition, something that makes a 5G network become another connectivity dimension: user plane latency for Ultra-Reliable Low-Latency Communication (URLLC), that is, the ability to successfully deliver an application layer packet/message from the radio protocol L2/L3 service data unit (SDU) entering point to the radio protocol L2/L3 SDU entering point via the radio interface in both uplink and downlink directions.

In this context, neither device nor base station reception is restricted by discontinuous reception (DRX). In plain English, what this new low latency transport layer of communications means is the ability of a packet-data network connecting IoT devices to provide, without disruptions, a constant transmission in both downstream and upstream. This assures that all services within a 5G Mobile Edge Computing (MEC) network are transmitted successfully, in real-time, and end-to-end without delays, a fundamental service delivery for critical applications, especially those involving vehicles at speed. This fifth generation of mobile communication systems promises to off-load onto Edge tasks which require applications to stick to strict latency requirements and, you guessed it, it uses artificial intelligence algorithms in order to partition tasks into sub-tasks, offloading them to multiple nearby edge nodes (ENs), just like electrical grids self-manage their networks to ensure that electricity is always available. The 5G ecosystem will not only perform for the automotive industry and its autonomous vehicles but to an ecosystem of other tasks that range from smart factories, remote surgery at smart hospitals, and other real-time control of cyber physical systems in a real or virtual environment. In the next ten years, the concept of "smart" or "intelligent" will encompass much more than cities: it will signify that many components of manufacturing and machine tasks will be automated with safety and reliability. In addition to ultra-reliable low latency communications, the International Telecommunication Union (ITU)[10] has additionally classified the fifth-generation 5G spectrum into enhanced mobile broadband (eMBB) and massive machine-type communications (mMTC), which denotes the substantial upgrade to network design and architecture that 5G will derive to the ITC industries and the increase of time-critical tasks that

---

[10]The ITU is a United Nations agency founded in 1865 to facilitate international connectivity in communications networks that strives to improve access to ICTs to underserved communities worldwide. The ITU allocates global radio spectrum and satellite orbits as well as develops the technical standards that ensure networks and technologies seamlessly interconnect.

5G networks will be mandated to deliver in a world completely digitized and dependent on it for substantial societal activities. To put this into perspective, current 4G networks are merely delivering telephony, social media communications, the Internet, and the exponential rise of digital media streaming. This is the current heavy-lifting. With 5G networks, High-Resolution Imaging, the Internet of Things, Cloud and Edge services and autonomous vehicles will be added to the list, representing an exponential escalation of data communications and latency requirements that will not only require milliseconds – as in the case of Virtual Reality services – but also high performance computation on a permanent, always-on basis. Moreover, this world of massive interconnectivity and communications will require that energy saving solutions are built into the 5G networks in order to optimize power consumption in machine-to-machine communications (M2M). As the rise in telecommunications exponentially increases, so will the need to upgrade existing energy resources and performance. The future, and it is more and more clearly outlined each day as we progress in ITC, will be heavily influenced by our ability to store and manage electricity. Whichever burdens formerly kept telecommunications growing at linear speed, constantly but in manageable ways, 5G networks will help the ITC sector achieve its highest potentiality, igniting as a result in an unprecedented need to innovate the energy sector.

Currently, 5G deployment around the world begins to unfold the birth of super-fast digital hubs in major cities and some early-adopter mobile phone users are beginning to switch to 5G-enabled mobile phones. Nevertheless, the infrastructure of these fast connectivity hubs is getting entangled in geopolitics, with the banning of Huawei technology in one third of the world's GDP according to Bloomberg. As of December 12, 2019, Australia, New Zealand, Japan, Taiwan, and the United States decided to break their contracts with the Chinese manufacturer of mobiles and network technologies, phasing out the company's products within their mobile networks. In the United Kingdom, against contracts that the Cameron government had previously signed off with Huawei, the current

Johnson government reduced their involvement in British 5G networks to just 35% from the majority stake that had been originally agreed. The reasons alleged for this ban point toward the threat of illegal espionage on the part of the Chinese government, who could use backdoors within the Huawei technology to have access to foreign countries' data. At the end of 2020, the majority of the world countries remain on the fence with regards to continuing with their Huawei contracts as more evidence surfaces regarding any trespasses of national security. So far, and to the dismay and annoyance of the Trump administration, nothing concrete or potentially demonstrable in a court of Law regarding government espionage has been found out against Huawei or the Chinese government. Still, the pressure imposed by the United States in some countries has created a divide between national decisions and private sector attitudes favoring to remain neutral. Certain telecom operators have decided to choose other network vendors as a partner for their 5G networks (in July 2020 Telecom Italia excluded Huawei from a tender for 5G equipment for the core network it is preparing to build in Brazil and Italy; in Belgium, both Orange and Proximus chose Nokia this past October 2020), whereas in recent days others have even challenged their own governments of restricting free competition and trade, a claim disclosed by the CEO Ericsson of Sweden in an interview with the *Financial Times*.[11] In addition to these national security and trade concerns, the Huawei ban will have serious consequences on the costs of building 5G networks. On June 7, 2019, Reuters reported that in Europe, the ban of Huawei equipment would increase the costs of building the European Union 5G network to an extra 55 billion Euros (US$ 65 billion) and delay the technology by about 18 months, something that European telecom lobbies are beginning ascribe to put pressure on their governments' national security concerns and use the values of commercial innovation as a winning argument. A year later, the European Union has taken a calculated approach to the problem, since

---

[11]Reuters, November 18, 2020

EU countries have competitive 5G mobile technologies from Sweden's Ericsson, and Finland's Nokia as well as South Korea's Samsung, and thus, has remained neutral in its directives. Both France and Germany, though not directly or explicitly banning Huawei technologies from 4G and 5G networks, have concluded to tighten their security measures and facilitate that other vendors are offered parts of the networks' servicing, making it harder for Huawei to remain with any foothold in their countries over time. How does a political crisis come to affect the private sector and how can governments remain in agreement on trade and military alliances such as NATO? An example of this is how Norway is responding to this tricky situation: whereas the Norwegian government has explicitly expressed their refusal to block Huawei from building the country's 5G telecoms network, increasing NATO's internal concerns, Telenor, Norway's state-controlled telecommunications company, has picked up Sweden's Ericsson as their key technology provider for their 5G network, forcing Huawei out of their infrastructure after a decade of servicing.[12]

Espionage and manipulation of other countries' data is a daily activity performed by intelligence agencies at the request of their government. Let us be clear on history and de-emphasize the current finger-pointing toward China. The United States and Russia have their fair share of espionage conundrums that range from the NSA tapping of prime ministers' telephones in developed countries to the mangling of US presidential elections in 2016 via social media fake news. Perhaps, the only way to resolve the current situation is to force diplomatic relations to measure the pros and cons of sustaining a global ban and mistrust of Chinese technologies in order to encourage world governments to vouch for transparency, cooperation, and de-escalation of national sovereignty. Diplomacy and international relations have worked out to

---

[12]REUTERS "Norway will not ban Huawei from 5G mobile network: minister," September 26, 2019; "Norway's Telenor picks Ericsson for 5G, abandoning Huawei" December 13, 2019.

deliver the highest benefit to humanity in Cold War endeavors such as the space industry, where the space agencies had fought hard to fence off government pressures that forbade collaboration and trust with other nations. Progress in the 2020 decade, as a benefit to social welfare, must derive from each government's peace and collaboration efforts with other nations. 5G and its ecosystem of connected services are a milestone in humanity's efforts toward the creation of a better-off society, a safer world, and, hopefully, a more entwined civilization that will learn to control its politician's egos and personal affairs tighter, encouraging a governance of higher vision and hopes for human evolution.

## Summary

According to analysts, 2021 is the year in which smart city projects, of which connected vehicles are a fundamental component, will start rendering true value to municipalities and the people living within them. The notion that a vehicle is today a connected computerized system is a reality, and this is even before it begins to self-drive. Telecommunication infrastructure has derived one of the biggest socio-economic benefits to humanity, not just directly in terms of allowing us to reach out to each other better and faster but by creating prosperity in economic and societal fronts. Now that the networks can offer a wide range of connectivity spectrums, the splendor of the Internet of Things, of Edge-computed devices, and intelligent transportation and traffic management are technologies that will exponentially grow to transform and optimize sectors such as healthcare, supply of goods to cities, road safety, and the flourishing of inner city communities.

# CHAPTER 4

# On Brand

*There is no other consumer good as emotionally potent as a vehicle. Perhaps an astronomically priced designer handbag, but nothing as powerful as an identity, as joyful and empowering to use. How the brand promise of vehicles has evolved over centuries is testament to the human desire of connecting with innovation, to enjoy today what seemed futuristic in an advertisement. To call a car "your own" is to leap into the realm of sentimental value. How can a machine generate this much love and affection?*

## Preaching to the Converted

When cars first came to market they had to be "evangelized": "Dispense with a horse and save the expense, care, and anxiety of keeping it. To run a motor carriage costs about ½ cent a mile!" read the first car advertisement in the United States in 1889. Prices were not cheap, at a $1,000 for a manually built model which, in today's money, it would amount to a $25,500 mid-market car, but the advertisement preached the technological advancements of a carriage never seen before: speeds of between 3 to 20 miles an hour, definitely faster than a horse; the engineering of its hydrocarbon motor, simple yet powerful; the elegant finishings of its carriage, the suspension wire wheels, pneumatic tires, ball bearings, and "no animal odor." It was a product marketing sales strategy over the costs and underperformance of owning a horse. The first car ad praised the ownership of something innovative and exciting but also cost-effective.

© Inma Martínez 2021
I. Martínez, *The Future of the Automotive Industry,*
https://doi.org/10.1007/978-1-4842-7026-4_4

Placed in *Scientific American*, a publication that was covering the likes
of Einstein, the effectiveness of this ad generated twenty-one buyers[1]
of a Winton Motor Carriage at the end of the year. The CPA (Cost Per
Acquisition) of this single ad was definitely well worth the returns of circa
$21,000 of sales. Still, cars were products for the few wealthy individuals
of the 1920s roads who had to additionally hire a mechanic to operate and
service them regularly. The pre-depression car advertising sector was in
line with the swinging times: cars were advertised directly to consumers
as "the perfect way to make *an entrance* to a ball," a similar simile to
how stretched limousines came to signify a "crazy night on the town" in
the 1980s. As a vehicle for commercial purposes to chauffeur wealthy
people around town, the ads also sold a brand value based on their
technical conveniences, a product value that many current taxi drivers
saw in the hybrid models of today. Contrary to consumers who bought
Toyota Prius in the name of lowering their $CO_2$ emissions, taxi drivers
and chauffeurs bought them for the practicality of reducing their petrol
costs. The early cars were objects of desire, public displays of wealth, like
diamonds, houses, and racing horses. Countries like the United States
and the European nations were in the middle of their second industrial
revolutions. Advances in mining and metallurgy allowed architects
to design and build skyscrapers in New York and Chicago. In 1883 a
long-distance train, the Orient Express, connected the cities of London
and Constantinople and five years thereafter a young George Eastman
commercialized the first film camera, the Kodak No. 1, under an innovative
pricing model: an inexpensive camera that would require peripheral
products such as lenses and film rolls upon which the company would
build their profit margins. Entrepreneurial innovations, industrial power,
economic growth, and a blue-skies mentality that anything was possible

---

[1] If you care for the trivia, the first-time buyer of this model was an English-born
Mr. Robert Allison, a pioneer citizen of Port Carbon, Pennsylvania, whose many
inventions for the mining industry became universally used.

were the driving forces of these times of economic progress. Inventors and empresarios were well aware that, in addition to mass-production and international trade, a fundamental pillar of their success resided in creating strong and recognizable brands, not just advertising. Between 1860 and 1884, engine and propulsion inventions propelled former horse carriages into steam-powered machines that later on evolved toward internal combustion engines. Still, these carriages were not considered cars until a German engineer, Karl Benz, designed and built for purpose a first car model: the 1885 Benz Tri-Car, a three-wheeled vehicle that he and his family drove around becoming the first motorists on the roads and the first vehicle to be awarded a patent as a new wholesome industrial product: an automotive, designed to be a car, not a repurposed, former horse carriage. Cars were a sign of the times, a product of an era of exponential innovation. An enormous leap in transportation, a mobile phone emerging from a sea of pagers. When the acceleration of scientific and technological prowess forces the velocity of change from linear to exponential, there is no way that the marketing of a product is sustained solely on basic design features. One has to create a brand.

# Brand Alchemy

The concept of branding is often misunderstood with the physical representation of a logo. It is not about branding cattle or putting a stamp on a document. A brand is an architecture of philosophical concepts that range from making a promise to a consumer to disclosing a product's purpose, its ethical standpoint, and its place in the world. A brand is the soul of a product, the manifestation of a company's reputation and values that are felt throughout the customer journey and user experience of products and services. Brand architecture requires to construct pillars of customer value and emotional connection points that cohesively create a unique value proposition, a singular, identifiable experience of what a product signifies in our life and in society. To achieve the perfect branding of a product is to seamlessly merge its core values of functionality, usability, and convenience with aspirational values such as trust, loyalty, joy, and other emotional states that we may wish to instill into a product. Joy, sensorialism, modernity, and futurism are four abstract values that have come to form the baseline of every car brand across decades of launching new models and consolidating manufacturers' branding over their roster of vehicles. Every new car is a newborn infant into the world of driving. People look out for unique features, transcendental innovations, and *raison d'être*: why the world needs this particular vehicle, and for that, not just brand campaigns but other powerful tactics were put in motion in order to convince consumers that every new model was a paladin of innovation, a magnet of desire.

# Glamorama

The automotive industry created excitement and branding messages at automotive fairs and industry exhibitions since the early days. The yearly Geneva International Motor Show (GIMS) in March, an event attracting more than 600,000 visitors and 10,000 journalists, has also an economic impact on the canton of Geneva, turning an estimated $225 million in local income per year. Yet the world's largest motor show takes place some hundreds of miles up the Alpine roads in Frankfurt, Germany. The International Automobile Exhibition (IAA for its German nomenclature Internationale Automobil–Ausstellung), organized by the Verband der Automobilindustrie (VDA – Association of the German Automotive Industry) and scheduled by the Organisation Internationale des Constructeurs d'Automobiles is the oldest in the industry since 1897, when it was held in Berlin. Starting with eight automobiles on display and a motoring industry that was still an unknown universe of inventors and car makers who considered themselves pioneers in an emerging market, the IAA has grown over the years in number of attendees, reporting in 2015 almost 932,000 visitors to the show, a year that also saw a record growth in car sales worldwide. Strangely enough, the visitor figures began to descend in consecutive years, profoundly affected by factors such as the launch of new models online as a premiere platform, and the pressure put onto car manufacturers to develop EVs (electric vehicles),[2] which until recent years were very hard to sell to consumers. Autovista Group, a respected pricing data provider and risk assessor in the automobile industry, concluded that "the sudden requirement to develop models with the new technology is taking a large financial toll on some manufacturers and, therefore, money needs to be saved. Suddenly, lavish stands at European shows seem irrelevant." Exhibitions, it seems, are beginning to show a decline as a brand display platform, and in turn, car brands have been forced to work hard on their brand architectures. It is the same story that has reduced industry participation at major events such as the Mobile World Congress, with household names such as Apple, Google, and Nokia dropping from the sponsor masts and the thousands square feet exhibition halls.

# F1: The Greatest Show on Earth

Formula 1, a powerful stage to test engineering innovations and showcase constructors' brands has also undergone dramatic changes to its revenue models and brand purposes. There are five automotive brands among the ten teams on the grid that use the Formula 1 platform as a petri dish for future innovations as well as potent channels of brand awareness. Global car manufacturers Mercedes, Renault, Honda, and Ferrari spread themselves across their own and other teams in an effort to keep the likes of McLaren and Alfa Romeo afloat and allow their roster of drivers to

---

[2]Autovista Group IAA "Frankfurt loses visitors but are EVs to blame?" September 26, 2019

train and improve their potentiality across other teams. It is a very tight-knit industry of friendships and rivalries that keeps the myth of racing automotive brands alive and generates millions of dollars in sponsorships and local racing circuits revenues. It is fundamentally a privately owned marketing machine worth $2 billion in 2019 that sells TV rights on a global scale and a private club for constructors that invest millions of dollars to own and manage a team. But just like the automotive industry exhibition fares, the Formula 1 2020 season delivered an unexpected casualty: Honda announced that it was leaving Formula 1 after the 2021 season. Honda, a constructor that had been keen to carve a name in Formula 1 returned in 2015 to McLaren, inspired by the new breed of power units that focused on hybrid and energy recovery technology. A clashing of cultures between the British team and their Japanese partners forced Honda to leave McLaren and come in 2019 to the rescue of two star teams, Red Bull Racing and AlphaTauri (owned and operated by an energy drink from Austria) who, after 12 years of wins, mishaps, and decline, dropped Renault as their power unit supplier. Even though the 2020 season has been an extraordinary first year at the top of the pole positions, Honda's decision hinted at an elephant in the pit lane that no one dared to reveal: that the automobile industry was going through a "once-in-a-one-hundred-years period of great transformation" that required budget allocations to support initiatives such as achieving carbon neutrality by 2050 and, in Honda's specific case, to electrify two-thirds of its global automobile unit sales by 2030. The decision also delivered a shift in their branding message away from the values that the Formula 1 context delivered to their fans and customers: Honda's future objective was to provide "joy and freedom of mobility" and the creation of "a sustainable society where people can enjoy life." In a single blow, Honda was implying that it was leaving behind the petrolhead industry in favor of developing new carbon neutral technologies that would create a happier mobility future. If the branding pillars of this industry, conveyed to audiences within physical environments and sports that involve fossil fuel engines no longer cut it, how will the automotive industry convey their values and purpose to the world when the industry shifts to sustainable energy and business models?

# Brand Messaging: Aligned vs. Contrarian

In order to understand the magnitude of this question, we need to look back at how cars were branded in the twentieth century. The first cars developed for the world roads where either racing motor vehicles or oversized displays of wealth. In the United States the 1950s car brands that were embedded in the hearts and souls of consumers continued to proclaim vehicles that were big statements on the road, oversized sedans and saloons that catered to big consumerism and the baby-boomers. In the 10 years after the end of World War II, the number of cars sold in the United States doubled in size to 50 million. Big was not only better: it was the only way to be. Detroit manufacturers quickly understood that the brand values to create for their cars should be based on vehicles turned into status symbols, so they feverishly manufactured new models each year with radical design approaches that could be easily identified by consumers. Consumers would spot new models on the streets with ease and assess who was driving the latest model that year. This is Apple's Jedi mind trick with their iPhone brand: each year, a new model is launched with a significantly uplifted design and 90% of the features remaining almost the same. Hardcore iPhone fans flock to the stores to purchase the new iPhone in the series in order to prove that they are able to afford the latest model. It is a product marketing tactic that keeps many businesses floating over the myth of their brands. In 1945s Europe, on the other hand, a world war had flattened the economies and spirits of countries that had to rebuild their societies with sobriety. The innovative, flamboyant, speed-mad European car industry underwent a huge transformation when their vehicles were forced to be smaller, more compact, less of a show-off. In France, the government restructured the auto industry according to priorities that were politically motivated. Hefty taxes were imposed on the horsepower of the Grandes Routières deluxe vehicles, so, in order to survive, French car manufacturers were forced to cease construction of large engine vehicles in favor of smaller ones. Renault, Citroën, and Peugeot launched small vehicles[3] marketed to families as "the people's cars," a branding that Volkswagen in Germany had originally deployed for their VW Beetle, a car that was designed to be accessible to everyone. The pain and suffering of the European population could not be insulted with fancy extravagant vehicles parading in front of the ruins of cities that had been flattened in air raids. This is a pivotal move that highlights one of the most important concepts of branding: aligning the product with the sentiment of the consumer. It is what Nike did with their Colin Kapernik campaign. It is what all brands are eager to achieve: to connect to the souls of the people, to what they deeply care about.

There is also another bold technique in branding: going against the grain, that is, to enter a market disrupting the very branding pillars that sustain the success of sales. This is what Volkswagen achieved in 1959 for the launch of their Beetle in the North American market. They turned to a newly formed agency in Manhattan, Doyle Dane Bernbach, and entrusted their creatives with a mandate that defied all the odds. For starters, the Beetle was not only an antagonistic vehicle in size to the super stretched models built by American manufacturers, but a model built by Nazi Germany, something that Volkswagen had to erase from consumers' memories and perceptions at all costs. DDB produced and executed two campaigns around the themes of "Think Small" (1959), polarizing

---

[3]Likewise in Sweden Saab AB was founded in 1945 to produce a small car model that could embody the unassuming and practical spirit of Swedes. Two years later their Saab 92 went into production: the entire body was molded out of one piece of sheet metal which gave it a sleek, aerodynamic shape, and its British racing green color was a serendipitous yet cost-effective decision to use Saab's surplus of green paint from wartime production of airplanes, a thrifty cost-saving mentality that paired with the times across the whole of Europe.

consumers on car size, and "Lemon" (1961), bringing to American audiences one of Germany's best industrial traits: obsession with quality and perfection. In a market flooded with vehicles built in the conveyor belts of Detroit, here came a small jewel of a car that had been manually inspected for the highest quality as if it were a hand-made Ferrari coming out of the Maranello factory. Volkswagen not only achieved their sales targets: they created a brand that catapulted them to a superior orbit above all other competitors, a brand that created new mindsets and attitudes in consumers that not only made smaller cars acceptable but desirable. Steve Jobs used to be adamant at not selling consumers what they wanted, but that which they didn't even know they needed or dreamt of, an ability that only strong and trusted brands can achieve.

# Purpose: The New Twenty-First-Century Value

The brand architecture of such brands is built upon purpose, not just promises. Brand purpose is a firm contract to deliver a vision and a mission that disclose what values guide your company and products. The Volkswagen brand has consistently delivered dedication to excellence and imagination to create beloved vehicles like the Golf, the hottest hatchback that every person wanted to drive since the 1970s. Marketed worldwide across eight generations and various body configurations and country-specific nameplates, the Volkswagen Golf continues today to be awarded car of the year.[4] It is a compact car that has evolved to signify a rainbow of values according to its many customer segments, from style, and modernity, to the fun of taking curves at speed for young joyriders. It is a model that, over time, defied its sector and became its default buy. A

---

[4]Just the Golf GTI has received the "Car and Driver 10Best" award for the 12th consecutive year in 2018.

pinnacle of success after the Beetle, the Golf was the cherry on a cake that had taken decades of car engineering excellence. And then, in September 2015, the Volkswagen brand woke up to a nightmare that equally shocked the automotive industry and loyal customers alike: the dieselgate scandal,[5] one of the most disturbing events ever lived within the industry, which not only affected them but also other respected car manufacturers.[6]

# Branding for Redemption

Corporate fiascos abound in the world of business because companies, as I incessantly repeat when I teach at business schools, are made of people, of human capital that can either drive it to success or sink it in financial distress or brand cataclysm. This news was a shock throughout the automotive industry because Volkswagen was beloved and respected for its merits across decades of hard work and innovation. A catastrophe of these dimensions not only shakes up the balance sheet of companies, but directly one of the most vital intangible assets: their brands. How Volkswagen has managed to rebuild their reputation in just five years is a story of bravery and resolute power of self-belief: in as much as the involvement of certain people blemished the company's brand, many

---

[5]The findings revealed that management in certain car divisions had been aware of an intentionally programmed turbocharged direct injection (TDI) in diesel engines that activated their emissions controls only during laboratory emissions testing. The result affected the vehicles' nitrogen oxide (NOx) output and the default device allowed the models to meet US standards during regulatory testing, though in the real world they emitted up to 40 times more NOx, affecting about 11 million cars worldwide in model years from 2009 through to 2015.

[6]On February 23, 2018, BMW admits having erroneously equipped vehicles with a non-conforming emissions detector and calls back 11,700 vehicles; On January 10, 2019, Fiat Chrysler accepts to pay up to $515 million to different industry authorities which have accused them of manipulating more than 100,000 vehicles; On February 25, 2020, BMW pays an 8.5 million Euros fine and a penal investigation case for "fraud" is closed.

more Volkswagen employees excelled at their jobs, living the values of honest hard work and consumer-centric purpose on a daily basis. The success of this turnaround was not just proof of how committed the company was to repair the damages done to their brand but how this disaster bestowed Volkswagen with a new mason stone upon which they could rebuild their organization to achieve higher levels of excellence by changing everything, from how the inner workings of their entire organization should operate with optimized protocols, to how strategic decisions should be made and how transformational and innovative the company should aim to be. By placing a new leadership team that would steer the company with radical innovations delivered with purpose and vision, Volkswagen Group set out to redevelop strategies and regenerate the brands of products with a twenty-first-century mindset focused on digitalization. Thus, the 2020 Volkswagen Group is stronger and readier to address several of the industry's most advanced transformations – electric cars, connected vehicles, and intelligent automation – all within an organization that has reinvented itself on a global scale. In addition, the Volkswagen Group has opted to enter the mid-market of electric vehicles wiping out the sacred cow of car ownership. Their full electric vehicles, recently launched in September 2020, will be offered under a financial instrument, Volkswagen Lease&Care, that will cover every cost associated with car ownership – insurance, repairs, software upgrades – and its sharp asset depreciation effects,[7] a single monthly payment structured to respond to our driving needs. To flip the revenue model of a massive corporation intending to put in the market 500,000 battery electric vehicles (BEVs) in 2020 is not a dare, but a stroke of financial genius. Volkswagen will be leveraging what they excel at: designing and manufacturing really

---

[7]Industry data from Black Book, which tracks used-car pricing, assumes a loss of between 20% and 30% of purchase value in the first year of ownership, and between 15% and 18% from year two to six. Overall, in five years, cars lose 60% or more of their initial value.

good cars and taking decisions that make sense for their attributes and the reduced size of their share ownership, which allows them to execute product strategies faster than other automakers. Pricing will play an enormous role in convincing consumers that they can go full electric. This is a bold bet on the 2021 sales expected in this category. How does an enormous global automaker such as Volkswagen switch gears in such a radically different way? Senior management consulted for this book did not hesitate to answer: the diesel scandal forced the stakeholders not only to re-imagine their entire business and vehicle fleets but also its many revenue models. Committed to completely transforming themselves to respond to the future they soon found out what sustainable vehicles will mean to consumers in a digitized society and what barriers should be lowered for such a radical business model in an industry that is still trying to come to terms with consumer and societal changes. Car manufacturers not only deal with what government regulators demand from them when it comes to $CO_2$ emissions but from polarized consumer attitudes that range from lack of interest in vehicles as a means of personal transportation emerging from a generation of Centennials who do not wish to own cars, to Covid-19 times consumers that recently have become aware that the safest transportation in a pandemic is the cabin of their car. This is not just about making lemonade out of lemons, and there is no pun intended here thinking back at their "Lemon" 1960s campaign, but a visionary attitude. The 2020 Volkswagen Group brand is standing tall on its heritage assets – a global presence that spreads from urban centers to small villages where one can easily spot their cars at local dealerships, and its radical approaches that go against the grain of an industry which is very reluctant to dramatic shifts. In a world of big, fat cars, they brought us the Beetle. In a 2020 society entering the sharpest rise in digitization, a range of pure electric cars that steer smoothly and grip the road like we are used to under a financial arrangement that will allow low income drivers to benefit from future technologies and government subsidies for choosing zero emission vehicles.

# Shifting Gears

The brand landscape in 2021 is beginning to shift toward a future of rapid transformational innovations that require new brand approaches. Striking and innovative brand propositions are starting to emerge with dynamic and challenging new positionings. The most significant and polarized is Tesla, a disruptor in the full electric sector. The EV marketplace is a hard nut to crack when challenging consumer mindsets and attitudes that assume that electric cars do not have any soul, or passion, and abiding government pressures to simply stop powering cars with fossil fuels. In 2020 this is no longer the case, and it is thanks to what Tesla and a few others have achieved. Tesla prides itself on a brand that sought to create a better, more fun, and quicker experience for EVs. Overtime, competitors have smartened up and launched electric vehicles that not only do this, but deliver superior brand values, for example, customer assistance points a stone-throw away from one's home, or electric cars that do not compromise on luxury experiences.

These are infrastructure challenges that Tesla is trying to compensate by excelling at what they do best: over the air software upgrades to a car's individual CPU and a brand proposition that has chosen to build its allure within a niche vertical: a cult-following among men[8] who love technology and see in Elon Musk a twenty-first-century bad-ass Steve Jobs figure that breaks the rules and strives to deliver dreams over "unnecessary features of the past." Tesla is adamant to change its game plan. It is an ambivalent brand that equally delights and wows customers as well as disappoints and confuses others. For example, the Tesla car interiors are reminiscent of SpaceX's Dragon Crew cabin: sleek polar white seats against carbonized black panels and not much else. Impressive, yet exempt of any excess,

---

[8]Tesla is terrible at appealing to women, who represented less than a third of Tesla car sales in 2019, perhaps because Musk's bravado sometimes is a deterrent to certain audiences.

it is so minimalist that even the glove compartment has to be operated from the central touch-screen, resulting in a terribly mind-boggling user experience for the common consumer. Tesla has quirky features that deliver greatness yet discards the simple things that should be standard, like a car bonnet that closes itself via hydraulics just like any other vehicle. With Tesla, you have to push it down by your own efforts as if it was a 1972 Ford Pinto. Ambivalent brands suffer from a lack of empathy sometimes. "Me, or the highway" or "this is how we do it" brand attitudes that overwhelm mainstream consumers, or worse, deter sophisticated customers that dislike to be forced into situations that they consider inappropriate: the back seat in Tesla models does not have an armrest in the center. That is why you do not see too many Tesla chauffeur cars in wealthy countries like the UAE awaiting passengers at the first class arrivals at airports. If you are fancy enough to be driven in style by others, you want to sit comfortably in the back seat, resting your arms on a soft-leather support piece that also holds your courtesy designer water bottle.

Apparently this must be a thing of the past to Tesla's mind. I like to compare Tesla to PayPal or Google: engineered to perfection but lacking in charm, and human-centric design. Nevertheless, this is Tesla's brand standpoint and their conscious decision: the future is one where armrests are obsolete, so let us all get on with it. Sadly, this attitude has had negative consequences when the industry rates their cars in terms of how they service consumers' expectations.[9] Still, Tesla maintains a market leadership with 18% of global electric vehicle sales achieved by October 2020. It is riding the wave of being the first full electric vehicle household, yet getting there has been achieved not just with great engineering but also thanks to government supported initiatives, such as its incredible success in Norway, a country that now leads in EV sales and whose capital is the epicenter of the EV sector. It was in the Norwegian market where, in 2013,

---

[9]In the 2020 Consumer Reports Vehicle Reliability Study, Tesla reported second to last among 20 brands reviewed.

Tesla broke a record for number of cars sold in a month for a single model, of any kind of car, not just EVs.[10] Industry experts, and myself included, predict that household names such as Volkswagen, BMW, Renault, Mercedes-Benz, Volvo, Audi Hyundai, and new Chinese automaker BYD will go on the offensive as of 2021 with less expensive models, leveraging from heritage brand recognition and after-sales services, a terrain where Tesla is still struggling to build physical presence outside the United States. The power within the Tesla brand resides in its decision to hyperfocus on a concrete set of pillars: super-engineered, masculine, expensive (therefore attainable to only a few), futuristic EVs. When their full electric Cybertruck was introduced as a concept vehicle,[11] the brand went octanes up in the ratings awarded to its "rebel attitude." No, dear lady, be not mistaken: this is not your pickup truck to go buy groceries, but a boxy monster truck straight out of Mad Max, the brainchild of a four-some team of designers led by Franz von Holzhausen, an American car designer behind all Tesla models, the Jony Ive to Elon Musk, who honed his talent across big names such as Volkswagen, Ford, and Mazda. With the Cybertruck it is clear that Tesla has a "rad" attitude and it is proud of it. It is deviating from the norm with purpose, raising the flag of a young brand, extremely brave and visionary. It bets on a product design concept that forces you to jump out of your reality and catapults you into the future – whether you like it or not. It is not a brand for the masses but for those few that want to experience visionary attitudes that are bold and unapologetic. It has been very successful and it still has to grow into an established brand that can

---

[10]In 2016 in the United States, 0.9% of cars sold were electric. In Norway, it was 29%. By January 2017, it hiked up to 37% of all cars purchased in the country. Source: Vox.com

[11]There is a superb review of Cybertruck by car designer Frank Stephenson that hits the nail on its head: Cybertruck is only an early version of its potential as a futuristic car. Its first version is sadly a heavily pixelated model in a PacMan videogame of the 1980s. www.youtube.com/watch?v=CjPi6Cn4D5M&feature=youtu.be

achieve not just market share but also consolidation, loyalty, and new verticals. It has definitely brought fun and excitement to a sector that needed precisely those values in order to make it relevant today, putting the bee in the rest of the automakers' bonnets to be better, to think harder, and to come up with competitive attributes that could challenge them. Tesla came to market betting on a power unit that was hard to deliver and wowed audiences with its Silicon Valley approach to disruption: be bold enough to challenge the status quo of any industry, especially those that comfortably turn billions of dollars each year. Skateboarding was a picturesque and skillful amateur activity in Southern California and then came the acrobatic riding of empty pools by street urchins that turned everything on its head. Tesla did precisely that. It has been the wake-up call to the automotive industry, just like SpaceX has been to the space sector. Will it keep its cutting edge values when, in five years, at least five other auto giants step in to pour millions of electric vehicles in the market? Competing on price alone will be very hard. Will it remain a niche brand for the few? I think it will continue true to its core values because each step of the way, we will need a Tesla to bravely go where no one has gone before. I just wished they included women in their brand narratives.

## Changing Lanes

And now, the polar opposite to the Tesla story, a "rags to riches" brand: KIA[12] Motors. The South Korean brand had very humble beginnings as a manufacturer of Mazda models for Ford Motors in the mid-1980s. It was a qualified, cheaper labor force that also delivered quality, yet it was forced to survive building cars for others. In 1991 KIA began importing its own vehicles to Europe and in 1992 it opened four dealership shops

---

[12]According to Kia Motors, the name "Kia" derives from the Sino-Korean words ki ("to come out") and a (which stands for East Asia). KIA thus is roughly translated as "arise or come up out of Asia" or "rising out of East Asia."

in Portland, Oregon to meet and greet their first United States consumers with a wide range of small-, medium-sized hatchbacks and sedans that did minimally well but failed to match the yearly sales of established brands. A year after the Asian financial crisis of 1997, KIA Motors sold 51% ownership to South Korean rival Hyundai Motor Company in a fire sale of bankruptcy and underperformance in international markets. How did the company manage to build a trusted brand on a global scale in just two decades? By relentlessly competing in every vehicle vertical niche with aggressive pricing and quality that, over time, began to position KIA as a serious contender, not just a provider of affordable cars. The KIA brand has placed itself within values that attract a specific type of consumer: people who cannot afford to send their cars to the mechanic every other month. They need cars built to last at prices that won't break the bank. "Dependable" is one of the pillars of the 2020 KIA brand architecture. This does not imply that KIA cars are exempt from engine and system failures. No car manufacturer can deliver such a thing. What has helped KIA become the world's most "dependable" brand is a clever product marketing strategy: KIA warranty programs. They are the longest in years, and the ones with the widest of covers. Basically, if you buy a KIA, you can forget about repair costs for five years. This is how brands come to elevate whatever the product features, or the pricing, cannot achieve single-handedly. Today, the KIA Soul is widely rated as the most dependable small utility model in its class while the KIA brand trades as the most dependable on the market. Still, KIA is aggressively tackling other verticals, even the luxury category. In 2020 their KIA Telluride achieved rewarding reviews and worldwide acclaim for its design approaches – dual panoramic moonroof, what's not to love? – and investing in superior materials to enhance the quality experience. It is a straightforward brand architecture: you want to create a luxury driving experience for contextual environments of mundane daily experiences, like driving kids to sports practice or picking up your dry-cleaning. It is the response of Asian automakers to the "Chelsea tractors," the overpriced Jaguar-Land Rovers that wealthy people in this London borough drive around, barely making it through the narrow streets of Kensington, unnecessarily über-powered cars for city living, but emblems of one's fat bank accounts.

The magic of brand architecture is a cosmos of emotional values that deeply connects with consumers. Branding is not about taglines around simple concepts: "Everyday" (Toyota), "Fahrvergnugen"[13] (Volkswagen), or "Shift" (Nissan); wordy narratives: "Everything We Do is Driven By You"[14] (Ford); and promises that sometimes make you cringe: "It makes you feel like the man you are" (Buick), "The penalty of leadership"(Cadillac), "For boys who were always men"[15] (Volkswagen GTI), and I shall stop there for the sake of pity. The truth is that the auto industry has, over the years, come up with brand campaigns that, if it wasn't for the high quality of their products, they would have ostracized most consumers. This is why it is incredibly difficult for them to sell to women, who, as consumers, have been bestowed by FMCGs,[16] fashion, and luxury companies with brand campaigns that are so sophisticated, abstract, and aspirational that it takes real expertise to be able to break the code to their minds and hearts. According to statistical analysis of registered vehicles and insured private vehicles in the United Kingdom, women tend to buy hatchbacks with lots of features, instead of choosing cars for their space or performance. We are the number-one buyers of white goods, that is, high performance household electronics. We consider cars just another member of such a category. Women are practical. We don't love gadgets as precious

---

[13]German Fahrvergnügen, from fahren ("to drive") + Vergnügen ("pleasure") popularized in the United States by Volkswagen advertisements in the 1990s.

[14]A wordy tagline with an attempt to be playful to convey that Ford is consumer-centric.

[15]No, it is not a translation from Martian to English. It is an original copy of total nonsense. To think that brands pay agencies thousands of dollars to come up with startling copies such as these is baffling.

[16]Fast-Moving Consumer Goods companies (packaged goods of everyday use such as detergents, toiletries, food, cosmetics, beverages, and other consumables).

collectibles. We do not have identity problems because L'Oreal Group already took care of that with their "Because you are worth it" campaign. Furthermore, our budgets are limited because we are paid less than men, sometimes 20% less for the same job title.[17] A sensible woman will not throw her life savings into a one-basket, $70,000 car if she can zip around in a sexy Mini Cooper, a BMW Serie 1, or a Mercedes-Benz Clase A – all three for around $30,000, a price tag that is not a coincidence. Will women drivers lead the adoption of electric hatchbacks? My money is here. If we can save in energy, plus get government incentives and discounts, and still drive around in good looking, feature-rich vehicles, the roads are ours.

What are the brand pillars of the 2020 decade? Sustainable. Ergonomic. Digitally smart. Autonomously Safe. Economically versatile. Freemium priced. Solving my problems, supporting my life. Feeling REAL.

## Summary

The brand values of vehicles expand beyond their core functionalities or our needs for transportation. Assigning emotional values to a vehicle is to highlight its embodiment of innovation, of being testament to an era, a year, a societal trend that captured everyone's heart. Today, when electric vehicles are making inroads into our streets, a new consciousness is emerging: what we drive can help the planet heal, save us money, and provide us with a new sensorial experience of driving, one that is subdued, silent, and open to let the sounds of the world come in. Branding EVs will require a whole new bag of tricks from marketeers, and a bolder attitude by automakers. Who will capture the souls of consumers with outstanding brands?

---

[17]For more information, check www.payscale.com/data/gender-pay-gap

# PART II

# From Transportation to Solving People's Problems

# CHAPTER 5

# I.Am.Car

*When did we start thinking that our vehicles were representations of our lifestyles? The marketing of new vehicles constantly seeks to appeal to our emotions. Is this a Kool-Aid we cannot resist? Does buying your favorite car make you the person you want to be? How have the automakers and their tactics turned us into this bunch of brand-obsessed, mid-life crisis drivers?*

•

## Beyond Transportation

For some people, the notion of a car is one of utilitarian objectives: transportation of humans and goods. The birth of cities was also guided by this paradigm of moveable commerce: roads, harbors, railroads, and airports have turned cities like New York, Rotterdam, Singapore, and Hong Kong into hubs of trade and economic progress. Seville, the city from where the conquistadores sailed onto a New World, became a cosmopolitan metropolis of 1500s Europe upon their return. Christopher Columbus would parade in the Spanish court ornamented not just with jewels and riches, but also with exotic talking birds and spider monkeys perched on his shoulders. Tended to by Indian servants that wore feathers, tattoos, and a cinnamon skin never seen before by Europeans, he would make orchestrated entrances to royal and high-class gatherings like a spectacular cornucopia of wonders. Becoming the living proof of his triumph, he equally bemused and shocked the Spanish society of the time. The spices, the fruits, the flora, and the fauna of these remote lands spilled over the European continent transported by horse and carriage

© Inma Martínez 2021
I. Martínez, *The Future of the Automotive Industry*,
https://doi.org/10.1007/978-1-4842-7026-4_5

over the old Roman road grid, by sea, aboard merchant boats that crossed the stormy English Channel to arrive at the remote ports of London, Amsterdam, and Hamburg. Transport built our civilization in bigger ways than religion or the military ever did. The automotive industry firestarted many human settlements and the mindsets and attitudes that emerged from them. The independence of movement that one acquires with a car, the most expensive purchase after a real estate property, creates a new realm of identity. While wealth allows one to buy cars as valuable collectibles and to be chauffeured around town, there is something deeply personal when it comes to driving a car of one's own, connecting to a powerful machine, and propelling oneself into an open road. It is a physical experience like no other, and our brain imprints a memory that traverses the concept of the self onto a realm of self-awareness and self-identity for many people. We become one with our vehicles.

# From Functional to Aspirational

Driving requires legal permits to do so. The first driver's license to operate a vehicle in the public roads was issued in Prussian Germany in 1888 to Karl Benz, inventor of the modern car. By 1904 it became a mandatory document for all drivers, and all vehicles had to be registered with their local authorities. Today in some countries having a Driver's License may even increase the probability of being hired in a job application as it is seen as a "qualification," though having one is still a personal choice and not a mandatory ordinance because it is dependent on many societal factors. In rural areas, owning and driving a vehicle is a real necessity in order to have access to pretty much everything: from basic needs such as attending educational institutions, shopping for groceries, driving to work or to a doctor's appointment, to familial and social needs such as driving to relatives to check upon the elderly members or help friends and neighbors drive kids to activities. Even in big urban centers where municipalities have invested in building adequate public transportation, driving automotives is a sign of independence and a proof that, if we can afford it, we value freedom of movement. Contrarily, when you have no real functional need for a vehicle, owning a car is a form of self-indulgence, a status symbol. Some people will aspire for freedom to roam, so any vehicle will do as long as it has got four wheels. Others will save every cent in order to purchase the object of their desires. This vehicle will speak of who they consider themselves to be, or what levels of desirability they possess. When money is no object, consumers tend to develop abstract relationships with the objects that they acquire, going beyond the traditional usability, functionality, and convenience levels of product design. Above and beyond what the advertising industry has contributed to create with their marketing campaigns, this psychological realm has not gone unnoticed to the entertainment industry.

Car models have seamlessly embodied a certain set of aspirational values and personal traits that film characters brought to the screen. A classic example of this is the James Bond film franchise and their long standing relationship with British sports cars. Even though the Aston Martin DB5 has been featured in seven of the classic Bond films, from Sean Connery's *Goldfinger* (1964) to Daniel Craig's *Casino Royale* (2006) and *Spectre* (2015), Bond's first car was a niche roadster built in Warwickshire: the Sunbeam Alpine, a two-seater drophead coupé produced from 1953 until 1968. But Bond was not its first driver on the screen. For his classic film *To Catch A Thief* (1955), film maker Hitchcock put Cary Grant and Grace Kelly aboard and the winding roads of the French Riviera as a backdrop. The combination became an iconic moving picture of sexual chemistry, fast and furious car chases, and glamorous intrigue, ingredients that have peppered Bond films across its 25 titles and five decades. Cars in Bond films are written up characters with real script lines delivered via their engineering beauty, technological innovations and stunts. When Roger Moore's *The Spy Who Loved Me* required that Bond's car became a Swiss army knife of sorts, from submersible vehicle to surface-to-air missile launcher, Lewis Gilbert the director turned to an extremely futuristic model: the Lotus Esprit S1, another British model. Even Bond's enemies have driven British models, like the Jaguar XKR, driven in *Die Another Day* (2002) by Bond's arch enemy Zao, creating a visual feast on ice when both models, Bond's Aston Martin Vanquish and Zao's fight each other in a Polar landscape. When the film plots took the story to foreign sceneries, Bond drove local national models in order to entwine the geography to the visual narrative. In France, Bond drove classic French models. In *For Your Eyes Only* (1981) the Citroën 2CV and in *A View to a Kill* (1985) the Renault 11 as getaway cars that have become iconic scenes of the franchise; in *Diamonds Are Forever* (1971) Las Vegas and its city lights became the set for Bond's escape in a Ford Mustang Mach 1 that was turned on 2 wheels in order to go through an alley. In Japan, Sean Connery's six-feet two-inch height forced the production team to turn his Toyota 200 GT into a convertible in order to fit the tall Scot into his cabin. Bond has even driven German BMWs: the Z8 for *The World Is Not Enough* (1999) and the 750 IL for *Tomorrow Never Dies* (1997), a product placing that showcased an emerging technology at the time: controlling cars remotely, and the power of film production dollars. Cars being featured into a Bond film narrative were synonymous of a manliness, power, and sex appeal that matched their owner, 007 Bond. In a current marketing world embracing emergent societal segments such as *metrosexuals* – heterosexual men *obsessed* with grooming, fashion, shopping, and similar interests traditionally associated with women or gay men – the Bond film franchise has transferred the traditional testosterone masculinity to the emotional appeal of vehicles, while softening the Bond character who since *Casino Royale* (2006) has found true love – not just sex, suffered sadness and depression from the loss of beloved partners, been weak and vulnerable, and, in sum, revealed a range of emotions that no other Bond of previous films ever did, all while his cars continued to stay strong, potent, and unmoved by events, keeping the franchise's masculinity flag standing and the testosterone fires burning.

# Humanizing Vehicles

The second tier of personification and emotional charge of cars has been splendidly represented by the success that a revamped BBC program called *Top Gear* brought to the 2002 audiences. Not only did the second version of *Top Gear* grow to reach 350 million people worldwide, a Guinness Record for the highest viewers of a factual TV program, but it also established a new perspective on motoring programs thanks to the humanization of vehicles and the emotional banter of its trio of presenters. All three motoring journalists, the team comprised of *The Times* social raconteur Jeremy Clarkson, known for his outrageous stereotypical diatribes and unadulterated dislike of any car "irritably perfect, inevitably German"; James May, known as "Captain Slow" for his prudent driving style, yet a jovial prankster; and Richard "The Hamster" Hammond, a dare-devil driver who spectacularly crashed a couple of times to the horror of producers and audiences. The unscripted chemistry of the three presenters created a fundamental concept for TV viewers: your car defines you, or even further along, your car obsession is not only acceptable and embraced, but the proof that you have a pulse, so hail to all petrolheads. This was not a hobbyist program for those bemused by engine sizes and vehicle statistics, but for the everyday fan of speed, pranks, and overall irreverence. Cars became the new hot property in television, bypassing cookery and quiz shows to jump into new non-linear television viewing: Amazon Prime and Netflix. The secret sauce? Personification of human qualities, candor and unscripted scenarios. In *Top Gear*, national diatribes and stereotypes created a personification of motor vehicles which became not just incredibly beautiful and highly engineered things to drive, but came to signify the meaning of life and happiness, the pillar of friendships and terrific fun as well as the core element of impossible stunts and challenges.[1] Cars were also turned into national treasures and embodiments of oneself. In *Top Gear*, unlike any other motoring program, cars acquired "personified descriptions" when models were reviewed and lyrical descriptions such as "Volkswagen Lupo has a 'mad face,'" or "BMW Zs have the best brains in the business" became metaphorical descriptions of humanized vehicles that the show hosts delivered without any preoccupations of infuriating the marketing departments of car manufacturers. Personified cars were easier to relate to in order to create emotional connections between vehicles and humans, as well as incredible engagement with the program's worldwide audiences. *Top Gear* is now a hybrid program that still runs on both British television – at some point hosted by, please be seated, Matt LeBlanc of "Friends" – and Amazon Prime, where the Clarkson-May-Hammond trio moved to, changing the name to "Grand Tour." In this web streaming format the concept has been elevated to encompass a worldwide, full-on, crazy-challenges circus on the road. *Top Gear* and *Grand Tour* have created an image of a "bad boy, joyrider turned comedian" in the audience's minds and hearts. It is irreverent, it is out of order, stereotypical in its reviews, but it is undeniably human and entertaining. *Top Gear* would be the TV program that Jerry Seinfeld, Kramer, and George Costanza would have hosted if their fictional characters would have been allowed to pitch it to NBC. Instead, as early as 2012, before the Internet became a streaming platform, Jerry Seinfeld starred and directed a program with a surrealist, René Magritte–inspired

---

[1]"Lap the Nürburgring in less than 10 minutes... in a diesel" (Series Five, Episode Five); "Lap the Nürburgring in less than 9 minutes and 59 seconds... in a van" (Series Six, Episode Seven); "Driving World Champion Fernando Alonso's Formula 1 car" (Series Ten, Episode Eight) and Wacky Races-style feats such as "Roadies van challenge" (Series Eight, Episode Eight) where the team decided to test some vans by being roadies for The Who and "Reliant Robin Space Shuttle" (Series Nine, Episode Four) where Hammond and May tried to convert a 1992 Reliant Robin into a space shuttle.

title of "Comedians in Cars Getting Coffee." Welcome to the talk show inside a car in motion.

With this concept, Seinfeld is taking the personalization tidal wave to a new emotional realm for TV audiences: to make cars and diners contexts of life, banter, and drama. Under the talk-show format, Seinfeld has been interviewing internationally known comedians in two separate contexts where casual and personal conversations happen inside cars and over the simplest of the meals: a breakfast. Seinfled has leveraged the intimacy of these two surroundings, both contained, small spaces, in order to create a "fly-on-the-wall" effect for audiences. These tiny geographies deliver what a traditional TV set fails to create: intimacy, closeness, humanity, truth, honesty, vulnerability. The choice of car model is decided according to each comedian's personal brand and so, the alchemy of human and car is created: a 2005 Cadillac XLR is chosen to represent comedian Brian Regan (Series 10, Episode 5: "Are There Left Handed Spoons?") in an episode where "It's the automotive equivalent of a man who colors his eyebrows, wears shorts with a belt and tucks his shirt into his shorts" and a 1976 Triumph TR6 is chosen for being "another great square-jawed, bulldog British sports car" for guests Colin Quinn and Mario Joyner (Series 1, Episode 8: "I Hear Downton Abbey Is Pretty Good ..."). What turns CICGC into "clever and engaging TV" from an anthropological perspective is that, in the majority of cases, none of the guests react according to Seinfeld's plans. They either hate the car and/or the breakfast, or worse, the cars break down or the planned route is unachieved as conceived – President Obama's security refuses Seinfeld to drive the President out of the White House grounds while the drive with Seth Meyers in a 1973 Porsche 911 Carrera RS is performed under torrential rains that make their journey to the Roebling Tea Room, a Williamsburg, New York institution, a grey and ordinary encounter, as raw a non-fiction, reality TV as one can get. The spot where they park is dilapidated and surrounded by bad graffiti and urban signs of decay, something that equally scares and shocks a Jerry

Seinfeld who wonders, unedited, if their car will be stolen. The camera frames a close-up of a vulnerable white and sky blue Porsche 911 that one can almost see trembling with fear.

# Hooked on Cars

The chain of emotions that cars and driving can create in humans has been a field of study in the sociology departments of some universities. In a 2003 study by sociology professor Mimi Sheller, founding director of the New Mobilities Research and Policy Center at Drexel University in Philadelphia, entitled "Automotive Emotions: Feeling the Car,"[2] she remarks that (quote) "Car consumption is never simply about rational economic choices, but it is as much about aesthetic, emotional and sensory responses to driving, as well as patterns of kinship, sociability, habitation and work." (end quote). Cars have come to represent a wide range of emotional identifications of the self, from personal success and achievement, to superior taste in design and appreciation of technology and engineering, similar to that of knowing how to buy racing horses or owning pedigree dogs. Steve Jobs was obsessed with German cars, driving them around as the ultimate technology product at speed. When someone walks out of a car dealership driving the model that makes them feel happy, oxytocin, the chemical that is released when we have physical contact or experience happiness, friendship, or complete trust in someone, turns the experience into a nirvana. It is the substance that elevates our sense of total joy as we slam our car's door and hear that hermetic *woof*, and the smell of "brand-new-car" cabin inebriates our pituitary glands. We wrap our hands around that coveted steering wheel and drive off to a new future of joy and self-confidence, not just the streets of our town. Some people have been caught

---

[2]Automotive Emotions: Feeling the Car. October 2004. Theory Culture & Society 21(4-5):221-242 DOI: 10.1177/0263276404046068

on camera driving off dealerships in their brand new cars only to crash absentmindedly into oncoming traffic. It is the kinaesthetic dimensions of product interactions that humans experience when we use our senses to fully experience products. Our brain limbic system releases chemicals that alter our emotional estates. Just like ink imprints fonts onto a paper, limbic chemicals are the substances that our cognitive system deploys in order to turn information into tacit knowledge, events into memories, emotions, assumptions, understanding, and embodiment of what we perceive.

# Nationalistic Assets

Objects of desire, humanized machines, the ability of cars to instill a wide variety of emotions, has also been leveraged by both governments and car manufacturers to create nationalistic business attributes. Translating this to a socioeconomics context, Alpha nations have an obsession with not just defending their home market, but infusing national price in their vehicles. This has had beneficial effects not just for the automotive industries of each country, but also for the development of concepts such as "competitive advantage," a business approach that the car industry and its national protectionism developed together with the US department of trade in the early 1980s. A Ford chairman, Harold "Red" Poling, once challenged Japanese car manufacturers to "build cars where they sell them," and so Honda became the first Japanese manufacturer to own a car plant in North America, in Marysville, Ohio, in order to bypass import restrictions imposed by the protectionist laws of Reaganomics America. Nissan and Toyota swiftly moved in the same direction and the market was soon flooded with Japanese cars that threatened the local automotive industry. The government commissioned independent market research and economic analyses that would aid in creating a national framework to incentivize and reignite the American industry. In 1984 two independent consultants – Clayton Magleby Christensen,

an American academic and business strategist who later developed the
theory of "disruptive innovation," and Liam Fahey, creator of the concept
of "competitive intelligence" – published a report coining the concept
of asset-based "competitive advantage." This corporate strategy was
understood as the ability for companies to perform in superior ways
to their competitors in the same industry or market by creating unique
attributes and resources hard to replicate. What this concept envisioned
was a marketplace where companies competed based on intangible
assets such vision, execution, innovative business models, human capital,
leadership but also their ability to leverage from their geographical
location, partnerships, and alliances with third parties and their skill in
orchestrating clusters of mutual benefit. When the relentless strides of
the Japanese car manufacturers began to erode the local market share of
American car companies, the boards of these heritage institutions turned
to advertising: consumers needed to be reminded of their national duty:
to buy American products, to choose American cars over Japanese ones.
"Buy American" was not a 1980s marketing slogan, but a trade rhetoric
that dates back to 1930s America and the nation's efforts to promote US-
made goods and ease the Great Depression. It is a crutch rather than an
offensive attack, a kind of dodging the bullet instead of sharpening the eye
and shooting like a sniper on a rooftop. While the 1932 "Buy American"
was an ode to American-made goods and their quality, craftsmanship,
and innovation, the current campaigns to ignite consumers to buy
American-made products are just a rebuttal to fend China off the world
markets, just like the 1980s campaigns to shift the scales against Japanese
companies. If we incentivize consumers to buy a national product just
because it will reinvest in the local economy, we are not forcing product
makers to innovate, but buying them home insurance, the notion that
their local marketplace will be safe because consumers will buy guided
by national pride. It is what probably drove US and UK car manufacturers
to their demise. The future of industries cannot be built upon World War
II propaganda. This industry worth billions, embedded into our lives

and the creator of transcendental innovations that have in every decade been the drivers of industrial progress deserves brave entrepreneurs and visionaries. This is the DNA of the automotive industry of the 2020 decade: incredibly competitive, as per usual, yet in possession of iconoclastic visions for the future.

## Summary

If we are to take our vehicles to the next level, they need to evolve to become not just transportation aids, or identity objects for our ego. The cars of the future look more and more like contextual spaces of our lives in motion, the environment where we will do much more than drive. The 2030 vehicle will be a functional space, a sensorized cabin where we will spend time engaged in activities other than driving, safe, comfortable, entertained, cozy, relaxed, happy to be seated inside a box of wonders. We will be marketed the experience of travelling just like the 1970s airlines enticed people to try glamorous transatlantic flights. The future vehicle brands are bound to evolve toward points of value beyond the original, twentieth-century individualistic ideals.

# CHAPTER 6

# Second Home

*What is the user experience of a car? Inevitably, it is precisely inside its cabin, at the steering wheel, but also relaxing on the passenger seat. Car interiors are a massive selling point for vehicles and a space of competitive tactics from one manufacturer to another. Seating in your car, you experience what it is to drive a luxury vehicle. Closing the doors of a highly engineered vehicle, one where, up to the last trim, the designers have thrown themselves to provide you with a superior experience of comfort, control, and safety, is where you notice the price tag. Slamming it hard, the door hydraulics and closing mechanisms let out the air, and hermetically close up the cabin, isolating you from the world's noises, welcoming you into the sound-proof nirvana of luxury interiors. With the arrival of digital technology, you are the recipient of experiences that go further than just driving.*

## Letting Go

By the second half of 2019, a report by the Society of Motor Manufacturers and Traders (SMMT) presented at the International Automotive Summit, claimed that "already more than half of new cars sold are available with at least one semi-autonomous driving feature and the vast majority have some form of connected technology."[1] It has taken the auto industry almost 30 years to embed automation and digitalization across the

---

[1]"Connected and Autonomous Vehicles: Revolutionising Mobility in Society," 2019 *SMMT Report*, Introduction by Mike Hawes, CEO, page 2.

© Inma Martínez 2021
I. Martínez, *The Future of the Automotive Industry*,
https://doi.org/10.1007/978-1-4842-7020-4_6

wide spectrum of vehicles manufactured in all shapes, sizes, and prices. According to the report, the economic impact of autonomous vehicles "could be worth around £51 billion per year by 2030, with more than 320,000 jobs created, the majority outside automotive manufacturing." Autonomous vehicles will not only improve safety and reduce the number of serious accidents due to drivers' errors but ignite the resurgence of economic growth, something that cascades back to society across every aspect of how we live, work, and play, forging a more inclusive society for those who need to drive for their education, employment, or basic needs and poor public infrastructure, disability, or old age become huge impediments. Connected and Autonomous Vehicles (CAVs) deliver the highest value to these segments of the population precisely because their need for mobility is not a luxury but a necessity. Since 2010 the design paradigm discussed across all vehicle brands is one that ponders on one fact: if humans will not manually drive their vehicles, what exactly will they do inside their cabins? Before we get to that absolute scenario of completely autonomous experience, the auto industry has been evolving according to the diverse human interactions occurring inside vehicles, from places to have fun and enjoy music and sound-based entertainment to back seat screens that play video games and films to your children in long trips, while cellular connectivity allows us to navigate via satellite, take office conference calls, and telephone other people while we drive. Cruise speed controls have for decades reduced the fatigue of pressing the gas pedal. Sensor-based technologies have automated our windshield wipers when the first drops of rain fell unexpectedly, and alerted us when we absentmindedly crossed over the highway traffic lanes due to tiredness or, worse, doing other things rather than paying attention to the road. Those car cabins have grown to become spaces of human behavior and interactions that today, when connected and autonomous cars are being designed for the most digital future ever imagined, their spatial dimensions and seating arrangements will be a significant departure point from decades past.

# Safe Haven

First of all, let us review how much time we spend inside our vehicles. Before the pandemic took over the world, despite growing efforts in urban centers to improve their public transport and create cycle lanes, and besides the popularity of affordable, on-demand driver services like Uber, from October 2018 to March 2019[2] the average American increased by 12% the time that they spent inside their car per week (from 9 hours and 43 minutes per week, to 10 hours and 50 minutes per week), even when daily use decreased by 10%. This has been reported to be the effect of aleatory poor weather conditions, the state of the roads in terms of ad hoc construction closures, and other events that affect the data and the assumptions that one could draw from the statistics. It also points out toward something more specific: once inside a vehicle, people spend more time at the wheel, and if this is the case, their time must be efficiently used for other activities, or for that matter, adequately turned into pleasurable and convenient experiences. There is a third element to these assumptions: privacy and the freedom to arrive and depart without the constraints of transportation schedules. Furthermore, since Covid-19 has turned city transportation into ghost trains and buses, let us add health safety to the mix. The new mindset that arose in 2020 among the driving populations is that our cars are physical environments where we are safe from contracting the virus, so car sales have gone up in every country, something that the migration of city dwellers to the home counties has also increased. These are the contrasting scenarios that are emerging for vehicle cabins in the 2020 era: privacy, convenience, multitasking, and health safety.

---

[2]Daily Ride Index, conducted by global communications consultancy Ketchum.

# Bringing the Ergonomics

In the early days of the motor industry, the interiors of cars were designed by a different team to that of the chassis and the exterior structure. Initially, the emphasis was on ensuring that the instruments fitted into a panel, and thereafter that the seats, door trim[3] panels, headliner[4] and pillar[5] trims were embedded in the cabin design following ergonomics, that is, with the customer in mind, ensuring that the humans inside would be able to comfortably operate the vehicle and be safe in case of an accident. How cars are designed follows a three step protocol: the designs are firstly sketched, then transferred onto a digital model where crash, thermo, and physical dynamics will be imposed on the design and tested for performance and endurance. Lastly, a full-scale clay model will be built to appreciate the wholesomeness of the prototype, a practice that was first introduced in the 1930s by automobile designer Harley Earl, head of the General Motors styling studio (known initially as the Art and Color Section, and later as the Design and Styling Department). The design of a car interior is the equivalent to the user experience design of software applications: it has to be intuitive and easy to operate, offer comfort and support, be fitted with safety features, and enable the driver to control the vehicle 360 degrees.

In the 1980s, with the emergence of computer systems incorporated to vehicles, manufacturers went wild with their interiors designs. In 1970, for the Lancia Stratos Zero, a futuristic-looking vehicle representative of the era of space exploration and the worldwide TV syndication of the *Star Trek* series, Italian designers planted a vertical computer flat screen – almost identical to today's Tesla control panel, placed between

---

[3]Industry term to refer to materials used from paints, plastics, and fabric designs to leather grains and carpet textures.

[4]Car cabin inside roof or ceiling above the passengers.

[5]Roof support structure on either side of a vehicle's windshield.

the window and the steering wheel. The screen was not mounted onto the dashboard but literally perched there. Two years later, Maserati launched their own version of the future with their Boomerang model, a vehicle that still preserves today its futuristic flare yet came to market with interiors that were a complete concept mess of six gauges, three switches and two levers encrusted around the steering wheel interior panel like fake, giant precious stones onto a jewelry box. Probably thinking that this would be an earth-shattering innovation, the control panel came without a speedometer, showing only readings for the RPMs, the fuel, oil pressure, water temperature, and battery. This makes Tesla and their back seat armrest omission look like a design mistake rather than a targeted design approach. The Italian designers ruffled the feathers of the automobile industry with flare and complete self-belief in their authority over what was stylish and what was not. In their design paradigms, the drivers of the future cared not for speed limits, and probably this was true for the laid back attitude of Italian road patrols who in those years loved nothing more than watching a *bella machina* speeding on the autostrada toward Lake Cuomo. Italy may not be Germany and their unlimited speed Autobahns, but speed limits were suggestions rather than lawful impositions.[6] Lancia continued along the decade striking poses at automotive exhibitions with concept cars that turned heads and made industry analysts ponder about the fine lines between design genius and design fiascos. In 1978 a newer and even more daring design was born out

---

[6]In the 1990s the European Union tried to unify the speed limits across the various member countries and set the maximum velocity on a highway to 120 kilometers per hour (74 mph). In Italy, especially in the North, where affluence allows many people to own sport and luxury cars, driving at 150 km/h (or more) has always been the "normal" speed. Italians went through the roof and managed to force their government to set it at 130 km/h (81 mph). In that way, whoever was speeding, was just 20 km over the limit, and no one was going to make a fuss about it. Unfortunately, technology and cameras have ended this cultural analogue practice and I will suggest you stick to the speed limits if you drive in the country.

of their Tatooine workshops in a galaxy extremely remote to normalcy: the Lancia Sibilo. This model came with an exterior made to look like a Swiss milk chocolate bar and, how to explain it with simple words, a steering wheel with no spokes or cutouts but resembling a Roomba vacuum cleaner robot displaying three oversized orange round buttons in the center. The driver would push the Simon game-style buttons to operate the warning lights to signal left and right turns and other visual maneuvers, while horizontal grooves for the output sound of the speaker system would blast the music straight toward the driver's chest, rather than coming to the laterally positioned human ears from surrounding directions. The latter seems like a comical episode of "somebody forgot to design proper speakers on the door frames and, at the last minute, they were chucked onto the oversized steering wheel which apparently still had room left to take up more design nonsense." All other instruments were displayed on narrow screens randomly scattered over the dark coffee suede dashboard[7] instead of on the vertical mount of the front dashboard, probably to compensate for the visual kerfuffle of the steering wheel and to keep the driver's eyes on the road where it mattered.

Since then, and across the next four decades, the design extravaganza of car interiors has had remarkable models that deserve mentioning. The 1980 Citroën Karin, with its pyramidal exterior frame – aerodynamics, anyone? – and three-seater cabin where the driver sat in the middle position, slightly forward than the passengers, redefined the notion of design weirdness with a steering wheel – why is this always the pièce de résistance that is destroyed beyond recognition by car designers? The Citroën Karin steering wheel appeared to have missed a wedge at the bottom, as if someone had cut out a piece of a cake. At a closer look, something that looks like the number pad of a keyboard sits on the center

---

[7]If you want to appreciate images and further descriptions of this unique vehicle, go to http://oldconceptcars.com/1930-2004/lancia-stratos-sibilo-1978/ and have a blast.

of the steering wheel, with two extra buttons beneath and a speaker above it – yes, they still thought that this location was the best place to create a sound experience, even though human ears are positioned on the lateral sides of our bodies. Buttons defined this decade of iconoclastic car interiors, with the Lancia Medusa's (1980) typewriter-style steering wheel of multiple buttons and the upgrade to the Lancia Orca (1982) which took the buttons design fixation to unexpected levels, making the steering wheel and visual screens resemble a colorful supermarket cash register attached to a heart monitoring system. In an era of red buttons pushed on presidential desks able to create nuclear Armageddon, of hard buttons in boom boxes that weightlifters took to the beach, and multi-caller lines telephones on Wall Street desks that men in $2,000 suits pressed to buy, sell, or shout at minions, buttons came to be the sign of the times in the luxury concept cars. Rich, powerful men wanted buttons to push! The future was one of buttons everywhere, each button able to perform a task, or a vanity.

Video games also altered the design sense of 1980s cars' dashboards. The Volkswagen Orbit (1986) presented a sleek, fully digital dashboard fitted with a satellite navigation screen labelled "Infovision Processor," a reminder that computer processing units were beginning to be fitted to cars' systems, and a design concept similar to the ones in Star Wars' x-wing fighter planes – something that the Pontiac Pursuit (1987) took to a full fighter jet cabin design, changing the steering wheel to a two handle bars stick full of buttons. As technological advances made in-roads into the automotive industry, the dashboards today have become tactile surfaces, the interior cabins have wrapped the humans around and reversed the traditional location of items, like in the Citroën Hypnos (2008), where the headrests hang from the ceiling, the seats are triangle-shaped, and everything seems to form a zebra pattern that flows around the cabin in rainbow colors.

# The New Interiors

Aside from the concept cars, interiors offer an opportunity to redefine car instruments reading as well as ergonomic concepts around comfort, safety, and control. In the twenty-first century, automotive evolutions toward sustainability, connectivity, and automation are shaping and redefining the kind of interiors that we consider appropriate and enjoyable for our current age of digitization, influencing all aspects of design: from the materials used to the social interactions that we assume will happen within them. Car interiors in 2020 also reflect superior concepts around aesthetics, comfort, materials, and user friendliness. In the cabins of sophisticated and state of the art engineered vehicles it is only fitting that simple things such as providing heated and cooled cupholders, massaging seats and ambient-lighting packages would make car buyers decide between one car or another. Trims have also been upgraded to new materials such as wood, carbon-fiber, microsuede, and even industrial denim. Electric BMW i3s offer interior surfaces made of more than 80% renewable, natural, and recycled materials as a selling point for consumers conscious of environmental needs and the circular economy. Elements neglected in the past like adequate and versatile storage are now a customer need that automakers are addressing with new approaches. The new requirement in cargo storage is a "spacious" and "well lit" trunk designed to hold large as well as smaller items such as groceries, bottles of wine, and shopping bags carrying fragile objects. Some brands like KIA have fitted under-floor storage bins in their new models.

Originally, car storage was thought to be a place for travelling trunks and luggage. In today's society, it is defined by how the auto makers address the day to day needs of consumers, not the once in a year circumstances of driving to a vacation resort, moving into a college dorm, or putting all your belongings into your car and move into a friend's apartment when you break up a relationship. With the increased liberalization of driving duties, drivers will bring more personal items into the car cabin: work documents, laptops, reading materials. This means that car interiors will have to accommodate increased spatial square footage for storing items for non-driving activities. This has created the emergence of a new concept: "modularity," the flexibility of changing parts of the interiors according to the use intended to be made of them. Storage compartments, head and armrests will be exchanged when we change activity, just like baby chairs and toddler seats have been incorporated to backseats when families increased in number of members and these additional passengers grew in size. Certain car models have been offering modularity for their seating arrangements, increasing seat capacity according to the needs of their occupants. New entrants into the car industry are seeing opportunities in gaps not addressed by traditional manufacturers. Yanfeng,[8] a joint venture between Yanfeng Automotive Trim Systems Co., and Johnson Controls, an American-Irish maker of heating, ventilation, and air conditioning controls equipment, have created a slim storage unit that wedges in between your car seat and the central gearbox, connecting magnetically to the floor console in order to catch the typical objects that tend to disappear through such crevices: coins, mobile phones, lipsticks, sugar sachets for your drive-through coffee. It is quite a surprise that it has taken over a hundred years to come up with a solution to this problem. The central storage modules have also been used for other purposes, perhaps going as far as installing an espresso coffee maker, a *Handpresso*, in a Fiat 500 L model in 2012 branded by Lavazza, the Italian manufacturer of coffee products. The single-shot espresso maker had been invented at the INSEAD university campus in Paris in 2006 and was offered as a luxury product across various verticals of outdoors activities, from camping to boating, appealing to Millennials

---

[8]A 30-year global automotive supplier, focusing on interior, exterior, seating, cockpit electronics, and passive safety, it is a successful example of OEMs coming together to unite know-how and expand their businesses to service the automotive industry as cars evolve toward living spaces, not just transportation machines.

obsessed with high quality barista coffee. The relationship between Lavazza executive management and the Agnelli-Elkann family,[9] made it possible to add this feature to a car model that was delighting audiences with its retro looks and fiery modern engines.

Modularity, the ability to turn the car cabin into pretty much anything, from office to entertainment venue is evolving to more sophisticated concepts in the autonomous car future affecting other components of the interiors: seats. In 2015 various luxury brands explored the concept of a club-lounge seating style instead of the traditional row of seats, all facing the same direction, onward with the motion flow. If people were going to be driven by an automated vehicle, there was no need to keep the train wagon concept of seating rows. Instead, the new car interiors present an inviting seating arrangement that encourages passengers to converse and interact socially, as everyone in the cabin is able to look at each other, resulting in the ability to engage more deeply and meaningfully. Customizable controls screens are now merging with sensor-based materials in the dashboard, allowing driver and passengers to use haptic movements rather than pushing buttons or switches, a trend toward tactile experiences that will increase the ease of functions inside car cabins. Instead of just touch screens, some manufacturers are considering the possibilities of other materials that can be electrified and sensorized and respond as digital instruments.

---

[9]The Agnelli family is an Italian multi-industry business dynasty founded by Giovanni Agnelli, one of the original founders of Fiat motor company which became Italy's largest manufacturer. They are also primarily known for other activities in the automotive industry by investing in Ferrari (1969), Lancia (1969), Alfa Romeo (1986) and Chrysler, the latter acquired by Fiat after it filed for bankruptcy in 2009. John Elkann was chosen as heir to the family empire in 1997 by his grandfather Gianni Agnelli who died in 2003. Elkann chairs and controls the automaker Fiat Chrysler Automobiles (which owns the Abarth, Alfa Romeo, Chrysler, Dodge, Fiat, Fiat Professional, Jeep, Lancia, Maserati, Mopar, and Ram brands)

# #MeToo

Eventually, all the disregard for accurate and meaningful user experiences is now being addressed with true purpose since we do so much more than drive in our vehicles. Finally, women-driver needs are coming up in the priority list of automakers for the first time in a hundred years, even though it was a woman, Bertha Benz, Karl Benz's intrepid wife and his main financier in creating the first motor vehicles, who became the first driver to go for a historic 120-mile round trip on an early August morning in 1888. As early as the 1980s, Volvo began to think about the needs of women drivers and their physical differences in body shape, weight, and height to those of men, founding the Volvo Cars Female Customer Reference Group (FCRG) to focus mainly on in-car safety, with particular emphasis on child safety. This division recorded gender-based data about women's use of cars and noticed that by 2002, over half of Volvo car buyers were women in the United States. A year later, an all-women team developed a concept car that was unveiled at the Geneva Motor Show on March 2, 2004, pointing at the obvious two facts: that women purchase about 65% of cars and influence about 80% of all car sales, and yet, teams of men still made most of the decisions in the design, development, and production of a car. "Your Concept Car," a three-door coupé, came with features that puzzled men but delighted women at the exhibitions where it was shown across America: the windshield washer fluid valve was moved to the exterior to keep the hood non-existent – because the car would electronically notify the owner's chosen service center when maintenance was due; no gas cap – because who wants to touch a greasy, petrol-smelly object to put gas?; easy-clean paint; numerous exchangeable seat covers of various colors and materials (linen, leather, felt, etc.); compartments for handbags – thank you Jesus, somebody got it; gull-wing doors that facilitated the loading and unloading of larger items and children, which could also be programmed to automatically open on approach; computerized assistance for parallel parking; and improved sight lines for women's smaller heights, an element of bespoke adaptation that car dealerships could enhance by conducting a body scan of the female driver in order to set the vehicle's seat position.

The female input when building cars is not a twenty-first-century oddity. Women have driven vehicles since year one of the motoring industry, contributing to some vital features of car interior design, for example, the rear view mirror. British long-distance motorist Dorothy Levitt, who broke the record drive from London to Liverpool in just two days, wrote a women's driving handbook published in 1909, in which she advised women to travel with hairpins (remember, the early cars were open vehicles), chocolates, a gun,[10] and a handheld mirror for looking backward, the latter being picked up by vehicle manufacturers yet not giving her any credit. "Most automotive companies expressed continual surprise when they discovered evidence of women's economic power," says Katherine J. Parkin in *Women at the Wheel: A Century of Buying, Driving, and Fixing Cars*.[11] "Although automobile companies did occasionally seek out female consumers, their fundamental inclination was to ignore them." Not Volvo. "Your Concept Car" never went for production, but twenty-two of its features found their way into Volvo models, including the Park Assist Pilot. In April 2019, Volvo went further in their efforts to make the automobile industry more gender inclusive by launching their E.V.A. (Equal Vehicles for All) initiative. Volvo's 50 years of crash dynamics data were to be offered to all automakers in order to

---

[10]Women feel unsafe ten times more than men. One of the key safety features in cars, to automatically lock all doors, is used by the majority of women to prevent strangers opening the car doors at traffic lights. I once had a drunken man in New York East side climb onto my Golf and try to get inside through the roof window, which was open to allow fresh air. He wanted to steal my handbag, which was resting on the passenger seat. Let this serve as an example of the kind of assaults that we women suffer, and how much of a target we are of car thieves. Hopefully car manufacturers will come up with better personal safety features in the near future.

[11]University of Pennsylvania Press (October 9, 2017)

improve how women's bodies and that of youngsters and children can be made safer inside the car's cabin in the event of a collision or accident. It is the same trend that made Microsoft acquire GitHub, the world's largest repository of open source software, or the sharing of data and scientific innovations that pharmaceutical companies have shared with each other in 2020 in order to accelerate the development of Covid-19 vaccines. The future seems to be moving toward an all encompassing society where the welfare of all supersedes private interests. Volvo brought to light a fact that the automotive industry kept in the dark: that in 2019 many automakers still produced cars based on data from male crash test dummies and thus, women run a higher risk of getting injured in collisions than men. So car interiors in 2020 are now addressing that women are 47% more likely to be seriously injured in a car crash, and 71% more likely to be moderately injured due to factors such as height, weight, seatbelt usage, and crash intensity. Shockingly, women are also 17% more likely to die because of how a car interior is designed.[12]

# Sensorized Cabins

Above all these innovations and progressive ideas around how to design car interiors, which to a certain extent were expected of an industry driven toward safer and seamless experiences of comfort and delight,

---

[12]European Union, Horizon 2020 Project, "Gendered Innovations: How Gender Analysis Contributes to Research," Tuesday, December 10, 2013. This publication presents the results of a group of more than 60 experts. With concrete case studies, the report shows that gender differences, in terms of needs, behaviors, and attitudes, play an important role in research design/content, and hence, the societal relevance and quality of research outcomes. It also reveals that these differences may vary over time and across different sectors of society, thus requiring gender-specific analyses. The report provides tools and guidance that will be useful to researchers when preparing proposals for Horizon 2020. Further case studies from EU and international research can be found on the Gendered Innovations website, which was created in cooperation with Stanford University.

there is something else that will forever change our relationships with our vehicles and what we make of their service to our lives. And it will have nothing to do with transportation but with our health. The car seats that manufacturers are currently working on operate as sensor-enhanced instruments able to take biometric measurements from driver and passengers. Car seats are being trained with artificial intelligence algorithms to interpret positional data and adjust the supports in order to help us sleep better or wake us up to make us become more alert. Seats are becoming dynamic, automated systems that will contribute to a better travel experience.

The sensor-society unfolding this decade will bring vehicles that will become safe havens, not just transportation machines. They will come to us fitted with health monitoring systems, so we do not have to buy them or pay extra for them to our healthcare providers. They will help us evolve toward becoming individuals who learn to look after their health, both physical and mental, allowing us to do as much as we need to inside our cars, which will be safe environments that will comfort and serve us with technological advances never before seen. Second homes that will look after our needs.

## Summary

We love sitting in our cars because car interiors have become spaces of comfort, joy, and pleasure. They are stylish and offer human-centric design concepts, luxury trims, and are beginning to offer sensor-based features that will derive health benefits, higher safety features, and an overall sense of assurance that we are being looked after, served in our needs, preserved from harm. Vehicle interiors are also places where no one can come and disturb us. Sitting in traffic, your car cabin is a parapet against the mad, noisy outside world of busy cities, a shelter from a thunderous rain that pelts the roof of our vehicle. The in-car experience is a new emerging area of innovation that increasingly will transform itself into a digital, interactive space, as well as a context of interactions that will be less and less concerned with driving, and more attuned to other human activities.

# CHAPTER 7

# Automation

*Getting to the full self-driving future has been in the making since the early days of vehicle innovation. How close are we from Level 5 driving automation and the complete entrusting of safety features and transportation duties to our intelligent vehicles?*

## Vehicular Autonomy

Humankind has harbored throughout the times deep desires to fly and propel ourselves at speed because life is an energy of forward motion. Being in control of powerful machines that plough through the oceans and steering vehicles that lift the leaves on the road with fury is an exhilarating experience, a cinematic imprint in our minds that makes us feel free, alive, and deeply immersed in joyful ecstacy. When did we dream of giving up such power? In the 2004 film "I, Robot" Will Smith takes over his autonomous Audi RSQ to the shock and horror of his passenger, creating one of the most memorable car chases in film history. Will Smith plays a detective in 2035, just 15 years from now, slightly paranoid about trusting robots, something that today still continues to concern thousands of people who believe that machines possess a conscience of their own, a mental stereotype that no longer applies to 2020 automation.

The intelligent machines that we are programming to function autonomously are precisely what machines were built for: precision, scalability, prediction. We, biological beings, cannot deliver that, which

© Inma Martínez 2021
I. Martínez, *The Future of the Automotive Industry*,
https://doi.org/10.1007/978-1-4842-7026-4_7

is why vehicle automation was born out of the need to reduce traffic accidents, in spite of what Will Smith's detective Spooner may feel in 2035. Our quest to build autonomous vehicles started at a time when science and engineering were flourishing in creating technological advances that were equally used for good as well as for evil, just like artificial intelligence in the early days of the twenty-first century. In the pre-World War II era, GM presented a concept autonomous car at *Futurama*, a World Fair's exhibition held in New York, which this year celebrated its 100th anniversary. The vehicle was guided by radio-controlled electromagnetic fields generated by magnetized metal spikes embedded in the roads of a model set depicting the highways of the future, a sprawl of aerodynamic pathways that crisscrossed the urban center of an imagined future city reminiscent of the one that German film director Fritz Lang dreamt of in his expressionist science-fiction drama film *Metropolis* (1927). At the 2020 *Futurama* exhibition, GM presented "Futurama 2.0," a promotional video clip offering their "Zero Vision" for 2039 vehicle automation, a world with zero accidents, zero emissions, and zero traffic congestion. While the 1939 *Futurama* imagined a society of autonomous vehicles and urban centers that gently melted into the countryside via "future" 1960s high-speed roads – years before construction of the interstate highway system, it presented the footprint of what the automotive industry will thrive to become in the years to come, an industry focused on safety, comfort, speed, and economy, objectives that have been carried forth to the 2020 society. Today in 2020, the next generation of autonomous cars are no longer testing prototypes but vehicles that have already been promoted to certification. Why the majority of autonomous vehicles today are still kept under wraps is due to the need for autonomous traffic regulation, the upgrade of road grids to digitized edge computing pathways, and bringing the entire car park of world vehicles to digital homogeneity so that all vehicles on the roads have computerized situational awareness of each other.

# Electric Roads

Projects to deliver the necessary infrastructure for autonomous vehicles started in the late 1950s when the Radio Corporation of America (RCA), a major electronics manufacturer created in 1919, decided to run an experiment in Lincoln, Nebraska in October 1957. A 400-feet road track was magnetized to demonstrate how vehicles could be run autonomously, controlled by a traffic control system. Think "Scalextric"[1] without the rails. RCA envisioned a future of connected vehicles that would join "controlled highways" driven by a human who, at the push of a button, would let go of the vehicle and let the road take over. RCA and GM worked purposely to further develop this concept of an electrified road, with GM creating various vehicle models that would connect to the RCA magnetized roads which, in turn, would control their speed and steering. In 1958 the company tested a Chevrolet with a front-end featuring magnetized coils that would connect to the alternating current of a wire embedded in the road to steer the vehicle according to the road wire tracing, without intervention from the driver. Another concept model, the Firebird III, could also demonstrate that the RCA "smart highway" concept worked, but the costs associated with building wired roads made it impossible for municipalities to afford it.

No one saw the point of such expenditure when cars were being marketed as joyride toys that empowered consumers to feel in control. It was research and development at its best, but a visionary joint effort that was picked up by university research departments, in particular by Ohio State University's Communication and Control Systems Laboratory where two different professors, Dr. Robert L. Cosgriff and Professor Robert

---

[1]Scalextric is a brand of electrified slot cars and slot car racing sets which first appeared in the late 1950s, manufactured by the English firm Minimodels. The forerunner to Scalextric was Scalex, which its inventor Fred (B F) Francis first produced through the company Minimodels Ltd which he had founded in 1947.

Fenton, a car enthusiast teaching electrical engineering, continued the development of driverless cars activated and controlled by a magnetized road. Fenton and his team worked on a prototype vehicle where the fundamental features of driving – speed, steering, and braking – were controlled by electronics. Images of a Plymouth, fitted with a sign on the roof that read "DANGER. Experimental Vehicle" and an electrified panel attached to the front, show a professor Fenton and his students, all white helmets and fatigues, being driven along unfinished portions of Interstates 70 and 270 while busily dealing with circuit boards and other gadgets. At the time, the Bureau of Public Roads had agreed to allow experimentation on autonomous driving and four states – Ohio, Massachusetts, New York, and California – were able to purpose-build electrified roads for testing. In the United Kingdom, the Transport and Road Research Laboratory successfully tested the magnetized road system on a tiny test track in Crowthorne, Berkshire. A Conservative politician, Lord Hailsham, sat in the front of a Citroën DS, theatrically taking his hands off the wheel to read a newspaper for the photo opportunity that the media in attendance suggested. The driverless Citroën DS managed to keep straight at 80 mph (130 km/h), demonstrating the safety of the system and the possibility that up to 40% of road accidents could be prevented with cruise control systems. In subsequent years TRRL tested other "hip" vehicles of the times, like Mini Coopers, and even proceeded to build about nine miles of track underneath the M4 motorway but the government did not go ahead with any further funding. Other automotive OEMS like the Bendix Corporation built and tested similar systems where the main component, the electrified road, remained present in the concept, keeping the paradigm of the "road controlling the car" unchanged from what GM had achieved with the RCA. In Ohio, Professor Fenton and his research team continued to request support from local investors and powerful figures like then-Governor Jim Rhodes, who enthused by their passion and the idea of furthering innovation in the state, eventually came through to support the creation of an engineering lab dedicated to vehicle research and development. Today the Transportation Research Center (TRC) in East Liberty, a small community village of just 366 people in the Perry Township of Logan County, is the largest independent vehicle test facility and proving grounds in the United States. Operating 24/7 – with approximately 4,500 acres of road courses, wooded trails, a 7.5-mile (12.1 km) High-speed Oval Test Track, and 50-acre (20-hectare) Vehicle Dynamics Area, or "black lake," TRC also provides performance driving instruction. Its 30 years of experience and homebrewed talent has allowed it to form global partnerships, alliances, and relationships with companies and institutions dedicated to vehicle innovation, living proof that vision and purpose can make excellence happen anywhere.

# AI-Based Autonomy

Meanwhile, the development of automated vehicles had not gone unnoticed by the military who had harbored a keen interest in robotics since the 1960s. Encouraged to transform military forces with autonomous vehicles, the Pentagon created three funding agencies, of which DARPA, the Defense Advanced Research Projects Agency, became the go-to funding source for technological innovation at United States' universities, a first line of investment that many scientific and engineering startups continue to take in their early development days. DARPA went West and funded various robotics programs at Stanford University where in 1964 young engineers began to develop autonomous machines. *Shakey*, a 1966 robot developed at Stanford Research Institute, now SRI International, was the founding ground for what later became Stanford's first autonomous vehicle, a self-driving cart that took up to 1971 to get right and which encouraged other universities to develop similar programs into intelligent automated logic for autonomous vehicles. The breakthrough of this new paradigm of autonomous driving resided in that it did not require for the roads to be electrified but instead used reinforced machine learning and camera vision for navigation. How one trains machines how to learn is an incremental process of teaching them how to sense, that is, how to gather data that is relevant to what they need to do, how to understand the world around them and make decisions that are appropriate at each instant. The latter is the hardest part to get right because the outside world is full of "mirages" and objects that look alike or move alike, and thus, interpreting them correctly requires incredible amounts of data for the training, and the correct labelling of such data to create a mental state in the machines that does not mistake a lamp post for a traffic signal. This is why you continue to train Google's image recognition algorithm inside "Captcha," the anti-bot software that asks you to click on images of bridges, pedestrian crossings, bicycles, traffic lights, and other objects along… roads. Did you think that this was a coincidence? We are training their self-driving image-recognition intelligence.

As the 1980s and 1990s unfolded, both industry and governments were excited at the prospect of creating semi-autonomous features and funded research and development projects to test this. In Europe, the PROMETHEUS Project[2] received €749,000,000 (about $2.1 million of 2020) in funding from EUREKA, a European intergovernmental organization for the research and development of innovation and scientific endeavors, which funded road safety projects led by universities and automobile manufacturers. In Germany, a joint development project between the Bundeswehr[3] University of Munich, led by professor Ernst Dickmanns, and Mercedes-Benz, made significant advances in the development of vehicle autonomy via computational commands based on real-time evaluation of image sequences derived from sensory data. Due to the low computational power of computers in the 1980s, the team had to devise sophisticated approaches to train the machine intelligence software such as which were

---

[2]Program for a European Traffic of Highest Efficiency and Unprecedented Safety (1987–1995)

[3]Bundeswehr is the German version of the US DARPA.

the most relevant details of the total visual input that it received from the vision cameras. In 1994, near Charles-de-Gaulle airport in Paris, the twin autonomous cars built by the team merged into one of France's busiest motorways, Autoroute 1, a three-lane highway where the twin autonomous vehicles travelled 1,000 kilometers at an average speed of 130 km/h and changed lanes left and right as instructed by the team, overtaking other vehicles with pristine safety. A year later, encouraged by the continuous success of their testing, the team made the car drive autonomously from Munich to Copenhangen, in Denmark, using active computer vision and being commanded to execute overtaking maneuvers in a kind of traffic that usually reaches 180 km/h as German Autobahns have no speed limits. The success of this approach was solely based on computer vision[4] without satellite navigation (GPS), pioneering many hardware and software configurations that today are used in robotic vision.

Back in the United States, DARPA, inspired by the success at Stanford, continued to fund other projects outside of Silicon Valley. In 1984, at Carnegie Mellon University in Pittsburgh, DARPA funded the creation of the Navigation Laboratory (CMU Navlab), a research and development facility to be run by Carnegie Mellon's Robotics Institute and School of Computer Science, focusing on developing computer-controlled vehicles that would self-drive using computer vision technologies, machine intelligence, and other navigation technologies. At CMU NavLab, many of the autonomous innovations such as LiDAR (laser vision used to locate obstacles) were pioneered in all kinds of autonomous cars, vans, SUVs, and buses. Just like their European counterparts, NAvLab began in 1995 to test semi-autonomous vehicles in the open roads of the United States, testing what later on would become the basic components in autonomous vehicles: GPS sensors, gyroscopes and magnetometers, time and date GPS antennas to provide precision time and frequency reference to the

---

[4]Dynamic Vision for Perception and Control of Motion (2007) by Ernst D. Dickmanns, Springer.

host computer system, as well as peripheral data acquisition systems, proximity laser scanners, omni-directional cameras, laser line stripers, and CCD cameras that used geometry to detect curbs and obstacles. All these data sets fed an algorithmic program that had been trained to handle live data. Additional projects funded by DARPA at the University of Maryland, the Environmental Research Institute in Michigan, and even the private sector,[5] gave way to the creation and funding of the Autonomous Land Driven Vehicle (ALV).

Combining the technologies developed across the various projects, ALV became in 1987 the first autonomous vehicle to drive off-road using sensor-based autonomous navigation, covering arduous terrains, inclines, large rocks, and vegetation. Autonomous now meant "free to roam," unbound from the roads, a scientific advancement that the US Congress did not hesitate to facilitate the necessary bills for to further its scalability and commercialization. The Federal Highway Administration and a consortium of private companies and universities worked toward *Demo '97*, a testing event that took place in August 1997, turning one of the city of San Diego's Interstates, number 15 North, into a test track for "automated highway systems," an encouraging effort to use technology and connected vehicles to optimize traffic flow and increase road safety by turning roads into interactive, digital infrastructure reminiscent of the magnetized highways of RCA and GM. The event attracted all kinds of automotive stakeholders and showcased with success many sensor-based features, such as vehicle-to-vehicle communications and "platooning" – cars following close behind each other at high speed.

Unfortunately, budgetary restrictions at the U.S. Department of Transportation impeded the further development of the program and, as no private company or investor picked up the $90 million tab, the hope for autonomous vehicles roads went back to the universities and

---

[5]Martin Marietta – a supplier of aggregates and heavy building materials for roads, sidewalks, and foundations.

car manufacturing stakeholders' drawing boards. The same year in Italy, another project funded by EUREKA, involving two universities within 63 miles from each other, the University of Parma and the University of Pavia, also tested their cars on the open roads. Backed by Prometheus funding, they created the ARGO Project to develop road safety via vision algorithms. The team created computer architectures that would allow the vehicle to compute its relative position with respect to the highway lane where it was travelling, extract the geometry of the road, and localize ahead any other vehicle and potential obstacles obstructing its path – pedestrians included. A year later in 1998, from June 1 to June 6, the prototype vehicle, a modified Lancia Thema, was made to autonomously drive a 2,000-kilometer route that took it to Rome, Ancona, Ferrara, Turin, Florence, and back home to Parma–Pavia. The "Millemiglia in Automatico" test succeeded beyond expectations, running on automatic 94% of the time under different weather conditions and traffic situations, as well as hills, viaducts, and tunnels. Of interest to the authorities, the project also demonstrated that an autonomous vehicle could be built with "low cost" technologies, but governments still hesitated to give autonomous vehicles their full backing.

# Wacky Autonomous Races

This did not deter the automotive industry who felt very strongly about continuing putting funding and resources to its development, especially DARPA, who knew very well how transformative these new technologies were for the future of the military. In March 2004, DARPA, in an effort to reignite the enthusiasm, concocted an unprecedented challenge: an autonomous vehicle race covering 150 miles in the Mojave desert. The prize was a million US dollars intended to incentivize robotics and automation scientists in helping the military to automate one third of its military forces by 2015. No one made it to the finish line at the first

DARPA Grand Challenge, so the money prize was rolled over to the following year's race. Just like in Vegas, the desert has that gambling effect of upping the ante and the challengers went for it even more vigorously. The October 2005 DARPA Grand Challenge had a winner: the Stanford Racing Team, with *Stanley*, an autonomous vehicle created in cooperation with Volkswagen Electronics Research Laboratory (ERL). *Stanley*, a modified Volkswagen Tuareg, crossed the finish line just ten seconds ahead of the Carnegie Mellon team that had their two vehicles, *Sandstorm* and *H1ghlander*, qualify in second and third places. The DARPA stables took the podium yet the rivalry between the two university teams was not without drama: Carnegie Mellon's *H1ghlander* suffered all kinds of problems and lost the lead to *Stanley* by just seconds. Another team from New Orleans, barely managed to get to the race as their lab had been hit by hurricane *Katrina* just weeks before the event. Contrary to 2004, when all vehicles were plagued by disaster, the 2005 race managed to have all 23 participants but one surpass the 11.78 kilometers (7,32 miles) distance record by the best vehicle in the 2004 race, proof of the exponential innovation achieved by all teams in just twelve months. Five teams in the 2005 race managed to successfully complete the 212 kilometers (132 miles) course. Delighted with the success and promotional value of the 2005 race, DARPA organized a third challenge, planned for 2007, this time, taking the autonomous vehicles to deal with urban environments. The 2007 Urban Challenge took the teams to compete at an airforce base in Victorville, California, and, still puzzling many observers today, DARPA decided to allocate them into two tracks, A and B, awarding the teams in the A track a $1 million in funding – *never complain, never explain*. This time, CMU, partnering with GM, took the pole position 1 with *Boss*, a 2007 Chevy Tahoe, while Stanford's *Junior*, a Volkswagen Passat Wagon, came in 19 seconds behind its tail. The 2007 Urban Challenge propelled the development of intelligent programs and hardware configurations that were superior to those in previous races as the tracks were built into urban settings where other vehicles and objects had to be autonomously dealt with by the racers. It attracted the participation of other automobile manufacturers and OEMs and cemented their compromise toward further development of vehicle intelligence.[6]

The original engineers that participated in the autonomous vehicles challenges were literally pulled out of their military-funded teams by Bay Area companies such as Google, Tesla, and Uber that were eager to commercialize AVs and grab the top talent in the sector. In 2008, Sebastian Thrun, who moved to Google after previously heading the Stanford DARPA team, spent the year bringing on board other DARPA Challenge alumni, including Chris Urmson, Anthony Levandowski, and Mike Montemerlo to respond to an in-house challenge coming down straight from the office of Larry Page, who at the time was still Google's CEO: to drive autonomously 100,000 miles of public roads and navigate ten 100-mile courses along well-known California routes, including San Francisco's Lombard Street. Google kept all these developments quiet and did not start to fully develop their autonomous car program until 2009. The autonomous vehicles industry had exploded and manufacturers of automated vehicles were now selling to the mining industry, evolving from the previous autonomous vehicles that used to be deployed in hazardous areas, such as inspecting the Chernobyl nuclear aftermath in 1986. In December 2008, Alcan, the aluminium division of mining giant Rio Tinto, began testing the Komatsu Autonomous Haulage System – the world's first commercial autonomous mining haulage system – in the Pilbara iron ore mine in Western Australia, confirming the benefits in health, safety, and productivity that moving to autonomous modular mining systems derived for their operations. In November 2011, encouraged by the continuous good results, Rio Tinto expanded its fleet to driverless trucks.

---

[6]Since then, DARPA's funding has swerved toward challenging competitions focusing on humanoid robots (2012), subterranean exploration via autonomous systems (ongoing since 2017), and nanosatellites (ongoing since 2020).

# Taking It to the Streets

The 2010 decade saw the entire automotive industry developing AVs, but one thing still remained unmoved: regulation. Timidly and very cautiously, a growing number of municipalities in Germany and in the United States began to allow AV testing on their public roads. Every research team, no matter the country, was nailing the new driverless technologies. Google, as ultra-competitive as they come, and with tier 1 manpower backed by the company's hefty coffers to fund anything that would be needed, went out of its way to lobby with anyone that would allow them to test in real roads and traffic. States such as Nevada and California began to pass bills that permitted autonomous vehicles on the roads with specific number plates displaying a red frame and an infinity symbol to signify "the future." Still, these were baby steps and the heavy lifting of sorting roundabouts, no-signalled railroad crossings, or school zones were still out of reach. AVs were still required to have a driver and a passenger on board. By 2014, the cameras, radar, and other technologies developed in the DARPA challenges and other R&D programs to make vehicles keep themselves within their lanes, avoid collisions and keep their velocity adapted to the flow of traffic, began to merge with other sensor-based features that ensured driver's alertness, assisted when parking and enticed consumers to experience how autonomous features aided their driving and improved vehicle safety. Google X, the moonshots lab created in 2010 where the autonomous vehicle team worked, was itching to commercialize their own *Firefly* vehicles but, in October 2014, Tesla took the market by surprise announcing that their Model S would be commercialized with *Autopilot*, their autonomous driving system. Silicon Valley was a hot spot of top AV engineers being poached out of companies or AV pioneers leaving Google to found AV companies.[7] Tesla spent the whole of 2015 testing their Model S up and down the San Francisco to Seattle highway, aggressively building up its driverless intelligence. The marketplace was a hotbed of AV action, but companies continued to develop their innovations in the "regulatory darkness."

# Levels of Autonomy

SAE International, a professional body for the automotive industry founded by Henry Ford himself and others, published in 2014 a classification system of autonomous driving intended to act as a guide[8] to standardize the different ranges of vehicle autonomy. In 2016 the guide was reviewed and updated to encompass new modalities and technological advances. According to this guide, a Level 0 car is basically a vehicle with no self-automation, capable of informing the driver of certain circumstances, such as running out of gas, but completely unable to perform anything by itself. Level 1 is a vehicle fitted with automated technologies such as Adaptive Cruise Control (ACC), where the car reduces its speed if vehicles ahead are travelling at a slower velocity, Parking Assistance with automated steering, where you sit at the driver's seat and your car maneuvers itself to park – almost motionless on a given spot, not travelling at speed, and Lane Departure Warning System (LDWS), a mechanism that monitors the road lanes and detects if the vehicle is veering out of its lane when not overtaking other vehicles. This

---

[7]Udacity, Thurn's startup, is an online learning platform that offers, among other things, a course in programming autonomous cars. Over the next two years, other key engineers quit Google's program to found their own self-driving startups. Levandowski forms Otto, which is quickly purchased by ride-hailing company Uber. Urmson joins Tesla's veteran Sterling Anderson to found Aurora Innovation.

[8]Code J3016, Taxonomy and Definitions for Terms Related to On-Road Motor Vehicle Automated Driving Systems.

is an automated warning specific to speed and safety, and not the correct functioning of car features covered by Level 0. Japanese car manufacturers began offering lane-keeping support systems in 2001 while Citroën was the first to offer the feature in 2005 outside of Japan. A Level 2 autonomous vehicle can accelerate, brake, and steer by itself though the driver must keep track of the road and take over if needed. This was the level of most autonomous vehicles in the 2010s. Aboard a Level 3 car, the vehicle's autonomy allows it to have situational awareness and predict outcomes, taking over the driver in cruising circumstances, such as travelling on straight highways, but requiring the driver to be conscious enough to take over if needed. The only vehicle on the market with level 3 autonomous technology presently available to consumers is the Audi A8. As of July 2020, Tesla vehicles were not yet truly Level 3 AVs. Level 4 vehicles are just beginning to be tested in Oxford, England, as part of *Project Endeavour*, a government-backed initiative that kicked off in October 2020. A Level 4 car can drive by itself but not be able to do it in unpredictable environments. In a Level 4 vehicle the driver disengages from whatever occurs on the road. For the Oxford project, Oxbotica, an AV company, has fitted a small fleet of Ford Mondeos with their autonomous systems, driving on a nine-mile circuit from Oxford Parkway station to the city's main train station.

# Autonomous Vehicle vs. Transport Robot

A Level 5 vehicle handles itself in all kinds of conditions, taking all decisions without human intervention. No commercial AVs are on the market at this level of autonomy but this does not mean that they do not exist in controlled testing environments. Google's Waymo is testing its cute egg-shaped cars in Austin, Texas, pleasantly driving around neighborhoods and merging into traffic intersections, while the AI recognition system identifies and interprets correctly everything in its way. Zoox, another company testing Level 5 vehicles for autonomous urban transportation, is a relatively recent entrant to the market, but very powerful in their vision and execution. Founded in 2014 by an Australian designer who sought a Stanford engineer that could realize his vision of a future AV, Zoox is a serious contender. Acquired by Amazon in 2020, the company has designed and manufactured a futuristic fleet of AVs resembling golf carts inside an ergonomic body frame with no discernible front or back and toy-style, cross handle faucet wheels. Zoox does not retrofit cars like the rest of the AV manufacturers, but has birthed a unique design concept resembling a toy mini bus to charm up an innocent and friendly invitation to hop on board. Zoox's VH model can transport up to four passengers, seating them facing each other inside the cabin, potentially co-working, sharing a meal, or playing "I-spy" games.

Zoox does not build cars. It builds transportation robots, because if you no longer drive, the vehicle that you are sitting on becomes something else, and this is what the mental stiffness of the automotive industry is having a real traumatic time to accept. We may drive Level 5 cars and press a button and make the steering wheel disappear, our seat turn around, and the cabin dim its lights so that we can have a snooze when we decide not to drive at all, but how long will it take for the entire mindset of driving vehicles disappear from our consciousness? The case of the Uber driver in Tempe, Arizona who in 2017 ran into a pedestrian jay-crossing a major highway, is a clear example of why Level 4 automation will be an intensely debatable legal headache. The Uber driver was charged with negligent homicide because she had been watching an episode of The Voice on her cell phone, instead of watching the road, which she did not monitor for 34% of the time in order to immediately take control of the car if needed. Her car was travelling at a mere 39 miles per hour (63 km/h) but the vehicle's automated obstacle detection *did not work at all*. There is a petty debate as to the system failing to interpret that the obstacle was a human, but the reality is that human or not, a moving obstacle traversing the road should have made the car apply the brakes automatically to avoid collision. It could have been a Moose in Alaska. What would have happened then?

The Uber driver's family would have sued Uber because, when you collide with an animal that size, you die on impact. Ask anyone up in the Arctic nations. If Level 4 AVs are going to require a human driver for "safety," what good are they for autonomous driving? Humans react very tardy if their hands are not right on the wheel, or their foot is off the pedal, resting on the cabin floor. Level 4 AVs are not made to watch anything but the road, and even then, we may react too late if something goes wrong, especially something that is not supposed to be there in the first place, like a Moose, blinded by my headlights, or a woman at 9:58 at night, crossing traffic lanes carrying shopping bags over a bicycle, putting others at risk.

## Autonomous Legalese

These legal dichotomies are preventing most countries from taking autonomous driving seriously. As of 2019, Australia, Canada, China, Germany, New Zealand, the United Kingdom, and the United States were the only nations with government-level discussions around autonomous vehicles, even though self-driving cars are not yet deployed at scale. In 2020, the first regulations related to automated features have appeared in the European Union[9] and in Japan, where the Automate Lane Keeping System (ALKS) regulation has been defined for legal purposes and scheduled to be applied as of 2021, there is still a lack of comprehensive regulation around autonomous driving classifications. The efforts toward AV regulations that are fully comprehensive are affected by the low number of companies testing on the open roads. Google's Waymo has been testing driverless taxis in Phoenix, Arizona, and autonomous shuttles have been tested in France, the Netherlands, and other European countries, but this is not enough to fully showcase to governments what

---

[9]Regulation (EU) 2019/2144 is defined in 2019 and applies from 2022 in the European Union for automated vehicles and for fully automated vehicles.

works and what is needed to fully deploy this technology with assurance. Out of the aforementioned countries, the UK government has adopted an aggressive and "no nonsense" attitude toward automated vehicles, proposing to bring driverless taxis to British roads by 2021, something that perhaps, the Brexit departure will turn into a national flag of pride and revive a competitive posture in the British automotive industry. The British government believes that AVs can make transport safer, easier, and more accessible, but deploying Level 4 AV taxis on such short notice and with just one test case in Oxford and without appropriate regulation is putting the cart before the horse. Even in the United States, where so many autonomous vehicles' innovation, testing, and breakthroughs have taken place advancing this industry like no other, the lack of government support for an automated vehicle regulation at federal level is still an issue. As of 2020 the federal government does not have any regulatory framework in place that can normalize the inequality between states that have moved faster than others in passing local laws to allow AV testing on public roads. If this ad hoc situation continues, the federal government will have to step in sooner or later.

Whether we drive "driver assisted"[10] vehicles or full "driverless cars," connected and automated vehicles will have a profound beneficial contribution beyond reducing the number of accidents on the roads and urban centers. They will contribute to the unburdening of tasks and duties in humans, reducing the pressures on our mental abilities, augmenting the opportunities to rest more, not just "do more" inside our vehicles. With the

---

[10]As of November 2020, Honda won approval to manufacture Level 3 autonomous vehicles in Japan. The first of the series is to be a Honda Legend which will feature this new automation level as a "Traffic Jam Pilot." In addition, and here is the whole point, Japan Road Laws were also modified to allow the entry of this new vehicle capability. If current legal systems do not adapt themselves to the fact that, at times, vehicles will be conducting the driving, there is not much future in continuing manufacturing autonomous vehicles. The laws need also to evolve. https://techxplore.com/news/2020-11-honda-world-first-autonomous-car.html

arrival of automation to the act of driving, we are offered an opportunity to completely let go, allowing our brains to occupy themselves with more pleasurable matters, pruning synapses that belong to twentieth-century road scenarios, and freeing up brain space for new mindsets around the notion that we can travel in a small vehicle of our own and not drive it. Remember when you first drove a steering assisted wheel? Or tried a parking assisted vehicle? These were "before and after" moments that, for many people, represent a departure point in their attitudes toward what and what not they wish to handle in their cars. It is very likely that, in the next five years, the United Kingdom and many more countries will indeed have driverless vehicles on the roads but not driverless taxis in urban centers. Even when travelling at lower speeds, driving in city traffic is forced to sort all kinds of unexpected events, and perhaps, even when the vehicle autonomy classifications will be standardized across the globe, the civil laws will not be adapted soon enough and the authorities will look for a culprit or a legally responsible party in the case of an accident. Again, the slow uptake of autonomous vehicles will have more to do and be more dependent on the Law than on whatever the transport agencies of the world agree. It will be, nevertheless, one of the most transformational events in the long history of automotive, perhaps the last pillar of what we today recognize as cars.

## Summary

The full automation of transportation machines – whether they are aircrafts, vessels, or vehicles – has been in the making for decades. Whereas airplanes and boats have been cruising the skies and the seas on autopilot matter-of-factly, trying to make road vehicles do the same is intensely intricate because current roads are analogue and urban environments increase the risk of accidents because a whole other world of moving objects and pedestrians add themselves to the mix, something that does not happen in the air or at sea. In those contexts, offering auto-pilot was to simply set a trajectory, and fix the steering to that coordinate, a straight line on a radar map that would never encounter anything in its path. Navigational routes, whether they are in the skies or across the oceans, are pathways agreed by the different marine and airfare management organizations. Roads are free-will environments where humans move at speed. If we are training vehicles to recognize not just objects, but to predict what those objects are likely to do within a context, artificial intelligence can do that and more. The main issue preventing the appearance of self-driving cars on the streets and roads is the fact that these elements of the urban and road environments are inert, they are not connected to the vehicle's spatial awareness, and hence, put all the onus onto the vehicle to handle itself in an analogue world. Autonomous vehicles will be safer when smart traffic edge computerized assets are deployed and the road laws evolve to consider that driving will increasingly be performed by machines and not humans.

# CHAPTER 8

# Together in Electric Dreams

*To a 1980s person of my generation, "electric" meant all that was cool and exciting: synthesizers, techno music, fluorescent lights, and bold colors that would light up our hair and clothes with luminescence to the shock and horror of the establishment. When the CO2 emissions began to put pressure on the automotive industry, electric was simply that, powered by electricity, and in the world of vehicles, to drive an electric car became the epitome of social responsibility. If driving an early-days hybrid was as exciting as driving a clothes iron, electric cars had a tougher crowd than that, with their battery needs, their handling, and that "cancelled-out" noise that turned the experience into a weird, sensorially deprived mishap. What happened next is the most exciting news in the history of motoring: we are going back to the 1980s, and "electric" is beginning to sound trendy, avant-garde, and futuristic as we make electric vehicles the next big thing in the future of this industry.*

## Why Go Electric Now?

It is worth understanding the drivers of the switch to full electric of an industry that has built fossil fuel power units for over a century. The principal force pushing for electric transformation is the government

© Inma Martínez 2021
I. Martínez, *The Future of the Automotive Industry*,
https://doi.org/10.1007/978-1-4842-7026-4_8

directives toward a CO2 neutral society. The European Union and China[1] are the two leading economic powers pushing for a green energy auto industry. The European Union, in specific, aims to reduce its CO2 emissions by 30% in the next 30 years, and considers the roll out of electric vehicles a big part of that plan. The year now is 2021 and, across most world's geographies, we are all aiming to achieve a full electric vehicle (EV) spectrum in the hydrocarbon-fuel-free society of the 2030–2040 decades. Most countries and states, like California, are said to be planning the ban on diesel car sales by 2040. Others, like the Netherlands and the United Kingdom, ten years earlier than that. Norway, a leading country in terms of e-mobility thanks to government financial incentives, is aiming to allow only low emission light vehicles, city buses, and commercial vans after 2025 and all new heavy commercial, 75% of new long-distance buses, and 50% of new lorries should be zero emission vehicles by 2030.[2] 2020 was a major target year, with Spain aiming for 2.5 million electric vehicles, Germany and India aiming for 1 million, Portugal targeting 750,000 and South Korea aiming for 200,000 electric cars. Country targets for 2030 include Finland's goal of 250,000 electric vehicles, Malaysia aiming for 100,000 electric cars, and South Africa's targeting a 20% share of electric cars. The switch to full electric is finally an objective that both governments and industry have agreed to achieve and there is no turning back. Nevertheless, and in the name of history, let us just state the obvious: before there were gasoline cars, there were electric cars. In fact, from 1823 to 1876, there were mostly electric cars, even some steam-powered ones.

---

[1]According to Bloomberg, China accounts for the largest share of global EV sales (predicted to account in 2025 for 54% of global sales) as it looks to reduce energy imports, clean up urban air quality, build its domestic auto industry, and attract manufacturing investment.

[2]New emission commitment for Norway for 2030 (www.regjeringen.no/en/dokumenter/meld.-st.-13-20142015/id2394579/)

# The Long, Winding Road Toward Electric Vehicles

In the early 1800s electric vehicles were developed simultaneously in Hungary, the Netherlands, as well as in the United States. In 1830s Scotland, inventor Robert Anderson developed the first crude electric carriage. All efforts rendered considerable success for this type of power unit toward the late 1880s, when French and US engineers led the way with the first true full electric vehicles and later, in 1898, a young Ferdinand Porsche, founder of the German sports car company by the same name, developed an electric car called the P1. These vehicles were powered by electric units that evolved over time from non-rechargeable primary cells, to galvanic cells, and lead-acid batteries which allowed for faster speeds and longer road trips.

At the turn of the twentieth century, steam powered 40% of American automobiles, while 38% ran on electricity, and just 22% on gasoline. The 1900s electric vehicles barely made a noticeable noise and had almost imperceptible vibrations. They did not smell like gasoline cars or required gear changes. Women favored electric cars because they did not need to be "cranked up" to get their power unit started or needed gears to be changed, something that required a strength that women did not have, so the independence that driving vehicles promised to consumers prevented most women to do so in gasoline vehicles. In the United States, Henry Ford, always quick to seize opportunities, leveraged his friendship with Thomas Edison and convinced him to work together in the creation of an affordable electric vehicle right from the start of his automotive venture in 1903. That year, as Ford incorporated his automobile company, Edison was already building nickel-iron car batteries. With the arrival of electricity to city homes by 1912, a total of 33,842 electric cars were registered, and the United States became the country where electric cars gained the most acceptance. Sadly, this year also marked the beginning of the end for EVs because they only suited urban use. At low speeds of between 15 to 20 miles (24–32 km) per hour, a short range of 30–40 miles (50–65 km) between charges, and their lengthy time required for recharging, there was one ultimate stone thrown over their fragile roof: their price was unattainable for most consumers. Even though Ford Motors produced gasoline cars, Henry Ford's mind and determination were vested in achieving electric cars that would be affordable by all segments of society. Newspapers like *The Wall Street Journal* and even the press as far as New Zealand were following Ford's endeavors into electric vehicles, wondering if he was pursuing a futile dream or the future of automotive after news went around the globe that Ford had acquired an electricity-generating plant in Niagara Falls, New York, as well as a new site adjacent to his current factory in Detroit dedicated solely to the production of the Edison–Ford electric vehicles. In an interview with the New York Times on January 11, 1914, Ford disclosed that the company was one year away from manufacturing an electric automobile, and that Edison himself was full-blast behind the project. "The problem so far," admitted Ford, "has been to build a storage battery of light weight which would operate for long distances without recharging." Holding the pressure throughout the entire year that the media put onto Ford to confirm when the electric cars would be available, the Edison–Ford project mainly succumbed to a human factor: the value of friendship over commercial issues. Ford would not admit that the vehicles would be fitted with batteries other than those built by his friend Edison. The problem with that was that Edison's nickel-iron batteries had very high internal resistance and were incapable of powering a vehicle in any circumstance. As Ford's gasoline cars increasingly required more of his attention and his accountants reminded him of the $1.5 million (almost $31.5 million today) in investment that the electric cars had amounted to without any sight of return, Ford began to withdraw from the project, perhaps convinced that his aims should be better deployed toward the money-making gasoline vehicles, perhaps ashamed to confront his friend Edison as to the banality of his nickel-iron batteries – Ford bought a total of 100,000 for the project – and complete and utter uselessness in powering vehicles. The "electrification" of gasoline cars, specially aiding the engine ignition and later on slowly resolving the problems that people found were unappealing, began to shift consumers away from electric cars. By 1920, America made gasoline cars the vehicles of choice. Even though some electric vehicles orders were kept up until the early 1940s, electric vehicles were relegated to fulfill niche markets, like milk delivery vehicles in Britain, or small vehicles for private use in post-war, fuel-depleted France.

# Storing Energy

The lesson to extract from the Edison–Ford drama is precisely what differentiates one electric car from another. Edison thought that any battery would do, whether one was aiming to power a kettle or a four-wheeled vehicle. The reality is entirely different. Vehicle batteries determine crucial factors such as the range of mileage that a vehicle can cover between charges, the retail price of the vehicle,[3] and the recharging speed and durability of the battery over time. A modern electric vehicle needs to account not just for the power required for mileage, but also to operate all other functions within the vehicle. Driving an electric vehicle at night and in cold or hot weather will demand increased battery power in order to drive with the headlights on and the climate control functioning for the entire duration of the trip. In this circumstance, the battery needs to be able to optimally supply for all eventualities and the various software systems that run the vehicle controls, from compressors and pumps to

---

[3]Around 2010, when EVs began to commercialize, battery packs cost around $1,000 per kilowatt hour. Lithium-ion battery pack prices fell 87% from 2010 to 2019. Today, for models such as the Tesla Model 3 or the Chevy Volt, the cost has gone down to around a volume-weighted average of around $150 per kilowatt hour.

auxiliary power and on-board modules. Car electronics today are a cosmos of systems powered by electricity especially with the increased automation of detection and motion cameras and sensors that are required to process images and data within milliseconds. In addition, perhaps the biggest differentiator between EV models, is speed of charge. The way that battery cells are developed today allow for faster charging times without compromising the cell's aging thanks to advanced nanomaterials in the chemistry of the cell and next generation electrodes that charge up around 200 kilowatts of power. The problem here is that in doing so they generate intense heat that needs to be cooled off within the battery pack. If the electric input is tricky, the output expected from EV batteries for sports cars is incredibly demanding. If an EV requires around 900 kilowatts of uninterrupted power, the output of this battery needs to offer around 1,800 continuous and 2,200 peak amps within a 90 kilowatt hour energy capacity, while dissipating the heat between the cells. So the challenge resides not just in designing an efficient input/output electric battery, but also an integrated, sensorized battery management system that can operate in difficult conditions and within a wide spectrum of temperature ranges. The temperature sensors within the battery pack ensure that if a cell overheats, it is immediately isolated so that the other cells around it are preserved, pumping liquid through the required areas to keep the temperature cool.

Designing the architecture of a battery pack is a challenge and precisely the area where the creativity and innovation of each EV contender and their OEMs are creating competitive attributes to ensure safety, robustness, and reliability. Tesla is the one EV manufacturer that has three of their cars in the longest range podium. At pole position number one is the Tesla Model S Long Range with a 379-mile range, retailing at a premium price of $78,490; at pole two is the Tesla Model 3 Long Range, with a 348-mile range and around the $65,000 mark, and at number three, the Tesla Model X Long Range with a 314-mile range and bordering the $115,000. I am not sure how far you need to travel each day to justify the expense, but if your needs are lesser and your pockets smaller, a Hyundai Kona Electric or a KIA-e-Niro will sort you out at around 278–282-mile ranges and under $50,000 each. Still, most people drive daily commutes of about 40 miles or less and the big automakers like Renault and Volkswagen Group targeted precisely that consumer group with affordable models such as the *Zoe* or the *e-up!* and electric *Golf.* It has taken the Renault *Zoe* just three generations to double its range to become the second most popular electric car in Europe, falling only behind the Tesla Model 3 in terms of sales.

Affordable full EVs do come with their own "not so great" issue: charging time. The new Zoe has gone up in power – 52kWh – good for a real-world range of 245 miles. The problem is, AC charging times have gone up as a result. If you are charging your Zoe from empty at home from a 7 kilowatt wall box, it will probably take the whole night, or about 9 hours and 25 minutes. Not bad if you plug it before going to sleep. You will wake up to a car ready to go. If you charge your Zoe using public chargers in your neighborhood, which tend to charge at around the 22 kilowatts, it will be ready in about 3 hours. If you are away and there is no proper EV charging unit nearby, you can even charge your Zoe using a normal electric socket, but it will take around 30 hours to charge it from empty to full battery. This is what deters most people from buying electric cars: the fear that one needs to be near proper charging infrastructure or the car will become a dead weight. Acknowledging this concern, Renault is also selling a 50 kilowatts DC charger for under $1,000 that can charge your Zoe to full in 1 hour and 10 minutes in case you do take long journeys frequently. The reality is that most people will be fine charging it at home at night or on the street every other day because their normal routines will be driving kids around to school and to extracurricular activities nearby, shopping for groceries, or driving to work within the city limits.

# Green Energy

Energy companies are heavily vested to the task of building EV charging infrastructure in countries where the government is also behind the expense. Enel X, the largest provider of demand response energy worldwide, is taking the lead in deploying circular economy initiatives, of which EV charging grids is one of the fastest growing endeavors. Creating partnerships with both car manufacturers, consumer-centric locations (shopping malls, parking lots, comfort areas along freeways and other spots convenient to EV drivers' charging needs) and energy operators to create network interoperability, the e-mobility sector is beginning to push forward the necessary milestones toward consumer switch to EVs and the commitment of local authorities to convert their public transportation fleets into full electric. There are even vehicle-2-grid technologies that enable electric cars to push energy back to the electricity grid and even power homes (vehicle-2-home) and buildings (vehicle-2-building), earning EV owners some money along the way. Leveraging from the original smart charging technologies, which unlike conventional chargers, can communicate with each other, your electric vehicle, and the grid, resulting in a cheaper, more energy-efficient and sustainable way to top up an EV, increasing or decreasing its charging power on demand, V2G claims to aid the widespread adoption of circular economy energy practices, recycling energy charges across all kinds of electrically powered objects via bidirectional chargers.

The V2G concept was conceived almost at the same time as the design of smart grids in the 2000s, when homes which generated their own energy supply via photovoltaic panels on the roof were said to also be able to send unused energy back to the grid. In practical terms, V2G is very nascent as a consumer behavior and it requires regulation as well consumer acceptance, but the decarbonization strategies of some countries are creating opportunities for both energy companies and electric vehicles manufacturers to jointly launch awareness programs to push V2G adoption forward. In the United Kingdom, two government agencies, the Department for Business, Energy and Industrial Strategy (BEIS) and the Office for Zero Emission Vehicles (OZEV), together with the government funding agency Innovate UK, have funded a V2G project created in partnership by Octopus Energy and Nissan Leaf. The Powerloop project turns electric vehicles into integrated energy assets connected to home and the grid that help support grid stability during times of peak demand, between the hours of 4 and 7 pm, and not just constantly extract energy from it. It is a consumer behavior worth testing: if you drive back home from work and your EV battery is almost full, why not feed it back to the grid and charge it back up from zero at an off-peak time, perhaps at midnight. How it works in economic ways for people who are willing to give it a go is relatively simple: you plug your EV to the grid before 6 pm and leave it there for a full cycle until 5 am the next morning. Using a mobile app, you tell Octopus Energy when you need your car recharged, and the grid will automatically charge or de-charge your car as planned. Complete 12 cycles within a month, and Octopus Energy will pay you a $50 cashback on your home electricity account. There is something uncanny about this new recycling lifestyle that may convince some people to go for an EV and dive into this type of cashback economy, a powerful incentive especially handy in times of economic constraints much more convincing than the notion of saving in petrol or reducing carbon emissions.

Now let us take this business model to its fruition and extreme. Imagine all EV owners obsessed with selling energy back to the grid. Eventually, and this is my own thesis, if every electric car owner was to off-set their home electricity consumption, energy operators will be forced to change their business models: people will net/net their energy consumption, and so, since they will produce energy as well as consume it in equal measures, their electricity bills may end up being zero, or – bad news for the energy companies – even negative, if they overproduce over their consumption, in which case, they would be credited in real, hard cash. I am not implying that this would be the case for a family of four, but very likely for single people who are barely home during the day, or night, if you tend to go out to dinner after work and you do not even cook at home or take your clothes to a dry-cleaners. Think about the amount of electricity that you do not consume from the grid because you pretty much come back home to just hit the sack and leave early the next morning. The new disruptive business model to deploy for energy companies is to sell you Energy as a Service (EaaS), where you eventually pay a subscription to have access to the electrical grid for your home and vehicle needs. Think Spotify: you do not own music, you simply pay to have access to it.

# Ctrl+Alt+Deleting Revenue Models

Cash will always be king in the consumer mindset if you are able to construct the right business models, which is what Volkswagen has done with their new fleet of EVs in 2020 with a leasing-only program: to sell you Transport as a Service. Unsurprisingly, and undeniably Germanic, there is a thoroughly calculated math applied here in order to supplant car ownership with car leasing, and it goes along the following lines: a typical car today, its body frame and chassis, has an average lifespan of about 300,000 km. The batteries in these EVs have a lifespan of 500,000 up to 1 million km. So not selling the cars but instead offering "transportation as a service" changes the whole finance paradigm. You just pay Volkswagen $500 or $800 a month, and you get your ID3 or ID4. You don't own it, Volkswagen keeps ownership of it, and every two years or so, Volkswagen gives you a new one. The benefit for you is that, for that monthly payment, you get everything covered: your maintenance and servicing, your car insurance, your electric fuel – because VW has set up a partnership with a company that sells renewable energy, and other packaged features that award you a full experience of what it is to have this vehicle in your life. One fee to one provider and you get all the transportation that you can eat. At the end of whatever the term is that you have agreed to lease the vehicle for, you give it back to Volkswagen, and you get a new one from them, keeping the same contractual agreement – and probably the remaining same battery of your previous contract. The car that they take back has a battery that maybe has made between 13,000 and 25,000 km, however much you drove in those 1 or 2 years, depending on your needs, but the battery is still at 99% of its original capacity, so they just take the top off, recycle that, put a new body frame on it, and sell it out again.

It is the biggest stunt ever seen in this 100-year-old industry: *long-distance* charged electric cars that one can lease over the lifetime of their batteries. It is the break point that is going to send adoption into a hockey stick curve, the before and after of full electric vehicles. Up until now most big manufacturers like Renault with the *Zoe*, or Nissan with the *Leaf*, have continued to sell their EVs for full ownership in the market. With Volkswagen and their lease-only model there is a sense of "being on a mission" to well and truly disrupt the EV market in all sense of the verb, even on car ownership costs. It is really brave of a household, heritage name such as VW to deviate from the norm with a finance shift such as this, but they seem to have grasped the nettle and none of the others have done it so far as seriously as Volkswagen has. Combined with government incentives, this could just be what the market needs to start growing EV adoption exponentially, especially in the European market, where European manufacturers have to fulfill the grams of $CO_2$ per kilometer that the European Union has set to come down from 120 in 2020 to an average of 95 grams across their entire fleet of vehicles in the market. According to Bloomberg's New Energy Finance Report entitled "Electric Vehicle Outlook 2020," it is believed that in 2040 some 60 million EVs are projected to be sold, the equivalent of 55% of the global light-duty vehicle market. Passenger EV sales jumped from 450,000 in 2015 to 2.1 million in 2019. Even though Covid-19 has put a spanner on many sectors and in particular, EV sales dropped in 2020 to just 1.7 million for the year, it is expected that by 2022 there will be over 500 different EV models available globally and that by 2025 around 8.5 million EVs are expected to be sold as battery prices fall, energy density improves, more charging infrastructure is built, and sales spread to new markets. It is not a twenty-first-century endeavor, though.

# Tesla: The Outlier

Unbeknown to most people, the state of California in the 1990s was enticing car manufacturers to build clean energy vehicles. From 1996 to 1999 the requirement to reduce CO2 emissions convinced General Motors to spin out a fully electric vehicle company: GM EV1. The trouble with being the first mover within the wave of radical transformation is that one needs a lot more synergies to make the business succeed. Even with the state incentives, it was up to consumers to buy electric vehicles and their green attitudes were still an anomaly, driven by a handful of people who had long cared for the environment with individual behaviors, like recycling, but that, though it began to spread toward a wider range of consumers who cared about their own individual *daily* emissions, not just the ones they generated when they travelled by air, it was unsustainable as a commercial endeavor. In these early days, the EV concept was still a research and development effort within the major automakers. For these multinationals, to turn EV production into full-fledged commercial products required a much bigger addressable market and there were some disheartening episodes[4] that determined exactly where the heart of the industry still resided: in the fast and familiar fossil fuel industry.

By 2003, the call of the clean energy future began to grow stronger on the mindsets of many consumers who started to display favorable attitudes toward clean vehicles. That year, when some people shopped around for a family vehicle they began to put on their requirement lists an unprecedented

---

[4]In 2003 General Motors recalled all its EV1 electric cars and then destroyed them, to the dismay of Elon Musk who tweeted about this as the "ignition" that prompted Tesla's co-founders to make EVs a reality.

item: CO2 emissions. If you were a family with two small kids, planning to go on trips every now and then, it also had to have decent booth space. So looking through all the cars that were available at the time, and filtering by emissions, one would get down to two cars: either the Ford Focus C-MAX, which at the time was 119 grams CO2 per km, or the Toyota Prius XW-20, which was a hybrid at 117 grams CO2 per km. Driving the C-Max was really impressive, and in some European countries about 3,500 euros cheaper than the Prius, but some people still bought the Prius, because driving a hybrid made them feel part of the new clean energy future. Full electric vehicles were still a rare breed, and multi-brand dealers had not yet registered a noticeable raise in customers buying cars not on price, but on emissions. The United Nations' Paris Agreement on climate change was still 10–13 years away into the future, and Greta Thunberg was just a month-old baby.

Still, in 2003 the EV challenge had not been completely abandoned but it was a flame that burned in the hearts of a few entrepreneurs and visionaries, not only driven by its promise of a greener Earth, but because of the tectonic challenge of shifting an entire mammoth industry like automotive away from its fossil fuel core. The thing about people like this, people who get a bee in their bonnet and cannot sleep at night thinking about transforming reality, is that some of them contain a special string of revolutionary genome within their DNA combined with an eager desire to solve massive challenges. The Tesla Motors' founders claim that seeing how GM "gave up" on EVs prompted them to challenge the move with disruptive approaches and their own take on how the automotive industry of the future should be, a bold statement that exemplifies why, if you really want to disrupt an industry, you need to bring on board people from the outside, people who will swing a bat at pillars that you and your colleagues would consider "untouchable," individuals who will not kowtow to any golden calves because they did not build such idols.

The 2003 Tesla co-founders, Martin Eberhard and Marc Tarpenning, were not mechanical engineers but computer science majors with proven experience in launching the first software versions of cloud computing.

145

Thus, they envisioned an automotive future that would combine not just car engineering, but clean energy propulsion, and proprietary software. In February 2004, when the one year old Tesla went to seek funding, Elon Musk went all out and took the lead, investing 86% of the $7.5 million round, and becoming chairman of the board. Musk was "on fire" in regard to Tesla but also deep to his knees in making SpaceX NASA-worthy, a task that was not exempt of failures and close-up competition from a deep-pocketed and equally obsessed about rockets Jeff Bezos. Musk was fighting tooth and nail to shake him off his tail, with innovative approaches to rocket engineering and propulsion, and a thin two years ahead advantage that was closing in as Bezos poured more and more personal funding into his Blue Origin rocket company.

Anyone would have passed on Tesla, but not Musk, who could not resist the immense attractiveness of Tesla's core value proposition, and the once in a lifetime opportunity to personally become an integral part of one of the most disruptive entrants into a sector which was shifting gears up toward its transformation and fullest potential. The remaining $1 million in the round was fulfilled by a small group of investors and the participation of Compass Technology Partners and SDL Ventures, a sign that this Series A round had investors still sitting on the bench, rather than enthusiastically throwing money at it, as it usually happens when the investment frenzy turns venture capitalists into investment lemmings or altruistic gamblers in the name of digital disruptions. Tesla's first round was Musk's act of faith, and investors committed funds in the name of friendship and somewhat FOMO[5] effect.

In May 2004, J.B. Straub, a key hire that would create the clean energy power unit, the third piece of the Tesla puzzle, joined as fifth employee and was later on recognized as co-founder. Tesla was a David in an industry of Goliaths that could shake around with one hand the miniscule EV startup challenging them under the radar. Tesla needed to grow more

---

[5]Fear of Missing Out

muscle, which meant to build momentum and put some models on the streets. Tesla did it Silicon Valley style: raising "in-your-face," pedigree money. Tesla's series B round was just $13 million covered by Valor Equity Partners, who put on the table bridge finance money while Musk bended the ears of every millionaire friend he had, enticing them to come and invest in the $40 million round that Tesla needed to truly put up a fight and squeeze itself inside an industry as the first fully committed EV manufacturer.

For Tesla, EVs were its core business. It was full electric or nothing at all. Three months thereafter, in May 2006, the C round was closed with Musk as co-lead investor and attracting the power names in the Valley: from Google co-founders Sergey Brin and Larry Page, former eBay President Jeff Skoll, and Hyatt heir Nick Pritzker, to the VC pedigree firms Draper Fisher Jurvetson, Capricorn Management, and The Bay Area Equity Fund managed by JP Morgan Chase, the banker of choice for the M&A action in the tech sector.[6] When investment banks participate in venture capital rounds, you can sniff the sweet smell of IPO in the air. This C round also had all previous investors reinvesting to avoid dilution, which meant that everyone was in, and no one was undecided as to its vision and competitive attributes.

By May 2008 Tesla managed to raise extra $40 million in debt, in a market that was slowly sinking into insolvency and the devaluation of all world financial markets. It is important to note the correlation between funding and innovation. Tesla would not have achieved its product roadmap if Musk would not have hustled the money and personally bet on its future. Not only did Musk use his name and fame, but also his design talent as well. While he was "possessed" by the emotional charge that seeing Tesla fulfill its destiny represented, Tesla's early days were still unknown to

---

[6]www.startupranking.com/startup/tesla/funding-rounds

most car enthusiasts. People in the know of electric cars first heard about Tesla around ten years ago, when Musk began to showcase the first models in development to car enthusiasts like Jay Leno, a Santa Monica neighbor of Musk, and getting the California governor at the time, environmental activist Arnold Schwarzenegger, to bring the electric car company's manufacturing to the formerly GM owned Fremont assembly factory in Bay Area[7] in order to build their sport sedan Model S, a $60,000 of 2008 full electric vehicle that was meant to achieve 225 miles on a full charge basis. The Palo Alto startup was no longer building prototypes in their garage. If they wanted to go public, they had to build cars like a proper automaker.

Being born in the *Mecca of Tech*, Silicon Valley, had its advantages not just for fundraising, but to entice the first Tesla customers. The Valley is not only where exorbitant money and tech fanaticism become the launch pads of disruption, but also the place where overnight fortunes are made by people under 30. The first buyers of Teslas were not the granola bar, eco-warriors concerned about $CO_2$ emissions. Instead, the early adopters were the titans of tech transformation, the emerging rulers of the digital society, of a future that turned the tables of complacency, the brethren that Musk himself once belonged to during his PayPal days. Driving a Tesla in the late 2000s was like having an iPhone the day after Steve Jobs pulled one out of his pocket in the 2007 Macworld stage. The future was attainable for some lucky people and the S model drove like a sports car should: with grip and gusto. Silicon Valley was also the geography where autonomous vehicles were being built, a target that didn't escape Musk's radar.

When in 2008 Musk became CEO of Tesla, moving its Palo Alto's headquarters to Hawthorne, Los Angeles, for the convenience of being able to run both Tesla and SpaceX at the same time, things were not

---

[7]Tesla initially sought to build its first assembly plant in Albuquerque, New Mexico, but the cost of building it from the ground up was ludicrous, and demand for Teslas was still stronger in the Valley, where the ignition that a young and thriving company like Tesla needed in its early days burned bright.

entirely hunky-dory at SpaceX. Some consecutive failed launches were putting the company at risk of losing investors' faith and the financial crisis had shrunk all venture investments into "wild" ideas, so challenging space rocket companies were considered moonshots, and no pun intended there. On a visit to Los Angeles, Jason Calacanis,[8] serial entrepreneur, angel investor, and technology influencer, got in touch with Musk, an old friend from when Calacanis used to run the *Silicon Alley Reporter*, a niche publication for Internet enthusiasts. Calacanis shared with Musk two principal traces: the serial entrepreneur DNA and a passion for sports cars. Calacanis was thus not just a fine-tuned spotter of emerging technologies, but an engaged angel investor, and someone renowned as a sound go-to person for advice and support. At their dinner, Musk admitted that Tesla had just three weeks of burn capital, a Silicon Valley metaphor to imply that it was three weeks away from running out of cash to run the company. Calacanis orchestrated that an anonymous billionaire would bail Tesla out with a loan and the promise of a new model S, a roadster model that Musk showed them in clay and that prompted Calacanis to order not just one, but two of them.[9] Calacanis became a Tesla evangelist, showcasing in 2009 a video one of his Teslas, the 16th Tesla ever built, a carbon fiber, fiery roadster in a Lamborghini bright orange with a Lotus

---

[8]In 2003 Calacanis's first venture Weblogs, Inc. turned him and his co-founders into the godfathers of profitable blogs, selling to Time Warner's America Online in October 2005 for $25–30 million and in 2007 getting Musk to invest as an angel in his next company Mahalo, a web directory (or human search engine) and Internet-based knowledge exchange (question and answer site).

[9]Calacanis claims that his two S models have serial numbers 00001 and 00073, and while researching for this book, I have found a video interview with him showing an S model with serial number 16 out of the first 100 series. Perhaps Calacanis bought a third S model. What matters is the enormous belief that he vested in Musk and how this paid off for Tesla, a favor like no other.

body look.[10] Calacanis was himself wearing a matching orange shirt and lovingly praised his Tesla's engineering glories of massive torque and light, distributed weight. Always the passionate fan, Calacanis dared to imply that one could even skip the thought of buying a Ferrari Enzo at a 200 thousand dollar ticket for one of the Tesla beauties, delving into geeky territory by providing tips on how to keep the battery at optimal performance, encouraging people to never charge the battery beyond the 80% capacity, and highlighting that even without that last twenty per cent, he could still get about a 190 mile range between charges, something that more than ten years ago was incredibly unattainable for most full electric cars in the making. *This*, the endorsement of one of the most respected members among the tech royalty and that of other big Silicon Valley apostles like Robert *Scobbleizer* Scobble and his evangelical tech blog posts, Flickr photos,[11] and videos, is what got Tesla the elevated status of "car of the future" within the tech community. If you made it as an entrepreneur in the Valley, you had to buy a Tesla. Pre-orders began to clog the Tesla books and to put pressure on its manufacturing capabilities, the black horse that the company always fought hard to dominate, the Achilles heel that industry analysts always brought up at the shareholders' meetings and company presentations, prompting Tesla at one point in 2018 to build an additional general assembly line tent on the parking lot of their Fremont factory in order to supply the massive demand for their

---

[10]Tesla's early days were dependent on its manufacturing partnerships for specific parts of the car platform, frame, and so forth. In a July 25, 2006, blog post entitled "Lotus Position," Martin Eberhard, Co-founder of Tesla Motors, explains to the Tesla fans, "one common thread that many of you raised: what exactly is Tesla's relationship with Lotus?(...) Much as I love cars, I am the first to admit that neither I, my co-founder, Marc Tarpenning, nor our original investor (and chairman of our board), Elon Musk, is an automotive engineer. We have quite a few excellent automotive engineers at Tesla now, but three years ago, we did not."

[11]There is historical footage of the famous night in Santa Monica when Musk showed Calacanis the first model S Tesla in Robert Scobble Flickr feed here: www.flickr.com/photos/scobleizer/2276158205/in/photostream/

Model 3 sedans, a target of 5,000 vehicles a week that Musk had promised their shareholders to fulfill.

Aside from having the best EV batteries on the market, Tesla has positioned itself above all other contenders thanks to a core attribute that challenges the traditional OEMs of car manufacturing and that precisely came from outside the auto industry. Designing and building a full EV is a pickle, not just in terms of the technology to manufacture and deliver them, but because the business model to make a profit out of them is full of challenges – if you sell an internal combustion engine vehicle today in 2021, you can expect to make about $36,000 off the vehicle over its lifetime in terms of parts, maintenance, and servicing. With an EV that drastically comes down to maybe $6,000, a considerable reduction in the traditional aftersales profits. So where do EV manufacturers make that extra money that they need to satisfy their shareholders? Software sales, and this is where Tesla is also making inroads that are challenging the status quo. Tom Raftery, Global VP, Futurist, and Innovation Evangelist at SAP, likes to call this the "iPhonification" effect: the hardware may be better in other handsets, but it is the user-friendly software that wins it for iPhones. Tesla is also the only car manufacturer that delivers software upgrades via OTA (over-the-air), just like you would update software versions on any computer or mobile phone.

The upgrades are no longer free of charge but sold as value-added features that enhance your driving experience further. Or so Tesla says. Under the radar, Tesla can also remove certain car features that came originally as factory settings, features that were included within the vehicle's original purchase price tag. According to Jalopnik, one of the best automotive blogs for car enthusiasts, buyers of second-hand Teslas began to notice the "disappearance" of certain car features such as the autopilot, which at time of purchase cost buyers a few extra thousands of dollars for the sake of driving a Level 4 self-driving car.[12] In a November 18, 2019 software upgrade, Tesla remotely removed all self-driving features, something that many Tesla owners have reported on the Tesla message boards.[13] This is a typical example of what Apple also does at times with their software upgrades and without prior disclosure to customers, for example, when Internet radio was removed from iTunes as a feature. How can this practice be legal if a customer has paid for specific features at the time of purchase? As per usual, regulators within the realm of consumer rights are asleep at the wheel where it comes to digital practices and their legality, but it beggars an answer from Tesla and Apple if they continue to still have a hand inside people's properties and move and remove features at will. In the meantime, and as it typically happens in software, there are some people out there who also know how to combat these abusive practices: welcome to jailbreaking your Tesla.

This practice, typical of cell phones, allows customers to have access to the root of an operating system in order to remove any limitations that the manufacturer may have imposed. In the early days, one had to have a friend or a friend of a friend to do this, but eventually jailbreaking shops began to cater for all needs. In the case of Tesla, owners wanting to get control of their software versions began to turn to Tesla repair specialists who would also see Tesla's removal of paid-for features as "theft" and be willing to help. This tide of surmounting discontent is not being addressed by Tesla, and the course of its current remote access practice is engrossing into a bigger problem if one adds to it the additional discontent that repairing Teslas is. If you are involved in an accident, repairing your Tesla is such a hefty ticket that most car insurers avoid it by writing the vehicle

---

[12]According to Jalopnik, the enhanced Autopilot and "Tesla's confusingly named Full Self Driving Capability" came together as options totaling $8,000 on a 2017 Tesla Model S https://jalopnik.com/tesla-remotely-removes-autopilot-features-from-customer-1841472617

[13]https://teslamotorsclub.com/tmc/threads/lost-fsd-earlier-this-week.156259/

off altogether. Tesla's car parts availability is scarce, as it is controlled by Tesla and not by third party mechanics. When a Tesla is written off by an insurer, Tesla stops supporting the vehicle by denying it access to the Supercharge network, even if you managed to fix the car and the car passed inspection by Tesla's own repair staff. They cancel this access remotely, without you having a say on the matter. An EV not being able to be charged is a piece of carbon fiber for the recycling yard, and this is angering more and more Tesla customers. When a software firm becomes paranoid to the point of overriding customers' control of their own devices, as it is the case of Apple, too, this becomes a point of friction that drives customers to other platforms, and in Tesla's case, face class action lawsuits or defection to EV competitors in the luxury EV segment like Lexus, Mercedes, Jaguar, or Audi. Musk must address the defection of loyal followers or this will be considered irresponsible. You cannot just annoy customers, treating them like pawns that you can lose in a game. Think Ben & Jerry's vs. the Pillsbury Doughy Man: the power of the collective, and worse if it is an "angry mob," can be detrimental to your brand and your sales.

# Everybody Is Doin It, So Why Don't Ya?

How can you explain the experience of driving a full electric vehicle? If you drive aware of the motion forces, you will notice the difference between a front engine, where you are being pulled, and a rear engine, where you are being propelled. In an EV, as the heavy battery occupies the gravity center of the vehicle, you feel as if the motion force is beneath you, like when you ride a motorcycle, but by placing the actual engine slightly to the rear or the front, depending on the manufacturer, and strengthening the suspension of the wheels, the EV delivers a playful feel of "true driving." The car feels connected to the road, even though you do not hear a sound as it glides over the asphalt, thanks to the position of the engine, which still gives you the sensation of driving a sports car or a sedan, so that you do not feel like you are driving an oil tanker at sea. EVs can go 0–60 miles/h in between 2.8 and 6.5 seconds to deliver the fun of quick speed and maneuverability, and they can reach overall top speeds of around 75–165 miles/hour, lower in the medium battery charge levels of 235 miles per charge so that joyriders do not burn the battery power too soon. Current EVs have an elegant handling that is surprisingly "grippy" when you take curves and, quite quickly, you begin to notice that you can actually enjoy not hearing a gasoline engine, and that it is quite pleasing. Will we miss the roar of a gasoline engine? Perhaps not.

## Summary

Switching an entire industry toward a new type of power unit will take decades, but all innovative efforts are now keenly focused on achieving the perfect driving experience as well as the precise price points and consumer journey to trigger a change to an electric vehicle. Let us not forget that driving cars is all about emotions, and that in order to enjoy the full EV experience, you need to learn to appease your mind where it comes to the "not enough charge left" paranoia that most of us felt in the early days of EV driving. Once you realize that your EV is quite capable of sustaining your daily driving needs, you start wondering why you did not own one sooner. Having said that, in order to fully enjoy the EV experience, you must have a home charging station, or the guarantee that a public one will be easy to get to whenever you need it. Once that downside is out of the equation, and if your driving is city, suburban, or within a reasonable distance, and you are not seeking speed thrills as your main motive for driving, switching to an EV in 2021 will change your life, your personal finances, and most importantly your conscience regarding helping push society toward a $CO_2$ neutral planet. Above and

beyond this, the human-centric experience, there are two big waves
of change: Transportation-as-a-Service, brought back by Volkswagen
Group and their bold move to lease and care for electric vehicle drivers
and their needs yet not selling a single car to them, but taking care of
their transportation needs for the rest of their driving lives, and Energy-
as-a-Service, that new business model that electricity companies will
test on consumers that are happy to switch to an electric vehicle and
begin to embed themselves into the energy equation of supplying
electric car battery charge back to the grid in order to offset their own
electricity consumption at home. The revolutionary effect of this will
soon bring forth a change of mindset in consumers – the notion that
electricity is a precious commodity that is finite and which we must
share with others if we have excess of it, as well as a new business model
that will charge people a subscription to an electric grid where they can
charge their vehicles but also sell back their excess capacity, a green
energy business model based on re-utilization of resources and circular
economy practices.

# PART III

# Your Car in the 2020 Decade

# CHAPTER 9

# Smart

*In my many appearances as an AI speaker, I always make the point of explaining the difference between "intelligence" and "cognition," the first one being a coefficient of mental ability, and the second, superior and infinitely more elegant, our ability to understand not just what is going on, or what something is, but to act upon it, to know what it is there to be done. In the world of automotive vehicles, this type of ability is turning cars into reactive, proactive, and ever more secure vehicles; adapted to fast-changing environments; and mindful of our biological needs to be safe from harm yet required to travel from A to B and then to C, for good measure. "Smart" has become the new black, and it is now not just about vehicles, but concerning everything around them, from roads, to streets, to urbanization of future cities.*

## Mindful, Thoughtful Vehicles

While assisted driving has revolutionized the early years of the twenty-first century, and fully autonomous vehicles will transform the entire automotive paradigm into a new dimensional concept of safe transportation, tranquil journeys, and the disengagement of driving duties, there is a previous layer of vehicle innovation that has been brewing for the last 15 years: being "smart." In this realm, your vehicle will not just be "intelligent," as in having a high degree of computational ability and AI-driven features, but it will be a "street smart" vehicle, in possession of good

© Inma Martínez 2021
I. Martínez, *The Future of the Automotive Industry*,
https://doi.org/10.1007/978-1-4842-7026-4_9

common sense to handle bad situations, displaying the necessary skills to react with precision and anticipation to everything that occurs around it, even inside of it.

At the moment, the fleet of connected vehicles and trucks that are available in the market simply take informational measurements of other vehicles' distance and speed, send and receive information about involvement in an accident, geolocation, and density of vehicle aggregation in a given stretch, but in the next five years, they will develop superior abilities in order to handle themselves accurately. To be "street smart" when we discuss humans is to possess enough emotional intelligence to acquire the necessary social skills needed to engage with others without conflict, integrating oneself into every contextual scenario. When we walk back home in the night, hearing our solitary steps on the pavement, we experience a "gut feeling" that warns us about potential dangers, like being mugged or assaulted, if you are a woman, a cognitive resort that signals that "something is odd here." Humans acquired this skill from the beginning of time and it is based on learning to develop our intuition. In the urban planning and demographic migration circles, we concluded that the city streets "make" you smart. Living in a big city center sculpts you into an "urbanite" and by the same token, if our vehicles will learn to function in this context, they will have to develop the same tacit knowledge to fit in, to belong to the streets. It is a "learned" skill, and thus, something that we can teach machines over time. In data analytics we train machines to operate in EQ environments via the detection of "anomalies" and, this being my profession, I can confess that anomalies are not just random meaningless events but the precursors, in many cases, of emerging trends, especially in behavioral analytics that observe and learn from humans' actions.

Because connected vehicles and street data will be live, extracted from our vehicles when they are in motion and from the connected road infrastructure, we will first train the AI systems in the Cloud, to get the modelling right, and correctly tag all data sets according to their true meaning and definitions within urban scenarios to assure data integrity. When we can prove the AI systems correctly understand the urban environment, we will embed them into both the vehicles and the infrastructure and from then on, they will operate within the Edge computing environment with live data, the data that only matters right there and then. The digital future is thus not just "intelligent" but aims to become "smart" because this is the realm where intelligent machines will learn to seamlessly integrate themselves into what we expect of them in "real life" scenarios. If we are aiming to develop autonomous vehicles that completely take over the driving, we will need to train them to be smart first so that they function within the reality of biological existence, where life is a context of expecting the unexpected, especially in urban centers, where machines and devices will have to deal with complexity.

The concept of "smart" is a telecommunications paradigm not subservient to vehicle energy efficiency or level of automation, like some marketing campaigns have implied. "Smart" is a concept around "vehicle connectedness" and how it will interpret contexts and make itself accountable within the "Smart Society" of 2025. In this reality, vehicles, homes, and cities will form an integrated ecosystem that Edge and Cloud computation will harmonize within a highly distributed software architecture. Since life is a continuous process of change and transformation, we must not just build computational intelligence in vehicles, but a "smart, socially adaptive" emotional intelligence (EQ) in all automated systems that will coordinate themselves to perform within the contexts of our day to day. An IQ trained AI system will teach a vehicle to detect the speed of the car in front and brake if this vehicle is reducing its speed. An EQ trained AI system will monitor how the car in front behaves, trying to "second-guess" what type of a driver is at the wheel by the number of abrupt brakes and hesitations in its speed, which will probably denote a learning driver, prone to put the foot on the brake too much. I personally try to spot if both their hands are on the steering wheel to verify if they are in control or if they drive their cars like a lawnmower on a lazy Sunday. Learning to be extra cautious in traffic is to train your brain EQ to drive in the real world and anticipate when people drive through stop signs or do U-turns in the middle of a two-way road. A soccer ball bouncing onto a street will probably be followed by a child, so you brake instantaneously. Nobody brakes for a plastic bag or a used coffee cup because your car will probably go over them without damage. How do you teach a vehicle to differentiate between the two objects and their contextual implications? By associations, by inference, by methods that we ourselves use in the real world. This is how we learn to really drive, and it is more an EQ context than a pure computational, mission control set.

# Seamlessly in Sync

How our cars and roads will become "smart" will be thanks to their connection to Vehicular Ad hoc NETworks (VANETs), a subcategory of traditional Mobile Ad hoc NETworks (MANETs). Vehicles and spots on the road pavement will become informational beacons. VANETs are designed to connect and manage vehicle travelling data and their communications with either each other in Vehicle-to-Vehicle (V2V) data exchanges, or with data transmission to Road-Side Units (RSUs) in Vehicle-to-Infrastructure (V2I) operations. VANETs facilitate that connected vehicles transmit information about their travelling destination and geolocation, their involvement in an accident, or their presence in a traffic jam so that other connected vehicles in the network make informed decisions, ambulances and public safety vehicles are dispatched to the scene where they are needed much faster, and the flow of oncoming traffic can be diverted to other routes before they congest the area. Road side units help local authorities and traffic management systems to account for vehicle data such as number of cars, vans, and trucks on their roads, their weight – which tells you how many tons of goods are delivered to cities on an annual basis, weather conditions, and other data sets useful when driving on the roads and traffic comes in and out of cities.

VANETs will most definitely be an upgrade from current Google Maps live traffic alerts. While Google's data streams derive from drivers' mobile phones and their geolocation within cellular networks, data that mobile operators are happy to share with Google for their Maps application, when smart vehicles connect to VANETS, these deliver real-time scheduling, an innovation used in cyber-physical systems (CPS) where computer-based algorithms control the interactions of deeply intertwined physical and software components. CPS allow vehicles connected to VANETS to operate on different spatial and temporal scales, exhibit multiple and distinct behavioral modalities, and interact with each other in ways that change with context, like how the flow of traffic is smoother or denser, slower or faster. CPS is the core component of what makes machines "smart" whether they are an electrical grid, autonomous automobile systems, medical monitoring devices, industrial control systems, robotics systems, or automatic pilot avionics. In electrical grids, if a node in the network falls, the flow of electricity supply bypasses the node or switches direction in the grid to still retain the voltage at correct levels and keeps delivering electricity to areas. In telecommunication networks of multiplex exchanges, and Internet Provider Services (IPSs), the packed data is veered onto another hub.

VANETs have also been designed to perform in *wireless multihop networks*, that is, next generation networks constrained by changing topology required to extend the service coverage area and to improve the wireless transmission capacity. This is the new backbone added to current VANETs to increase their ability to service the growing number of connected moving objects across wider areas and to handle physical communication systems variants – Single-Input Single-Output (SISO) and Multi-Input Multi-Output (MIMO). Vehicular communications within VANETs present increased connection dynamics due to different patterns of mobility, the effects of speed, vehicle density, radio broadcasting range, node rank, and connection duration. Think about how more complex this is for a cellular network that, up to now, only concerned itself with providing connectivity to a mobile phone but left device-to-device proximity communications to

Bluetooth. A VANET has to position all objects on the move and support the communications among them. What this next generation of networks brings with the addition of vehicular data communications is such degree of live information that traffic flow and safety are expected to allow cities to continue growing their active vehicles capacity and the access of goods and services into urban centers without collapsing the streets and locking free movement.

VANETs currently handle mobile communications and the streaming of large packets of entertainment data as well as V2V short and medium range communications, supporting short messages delivery, and minimizing latency in the communication link without increasing network costs. If we want VANETs to handle more complex Vehicle-2-Infrastructure communications, further investment will be required for the installation of advanced antennas to procure VANET/Cellular interoperability, and WiMAX (Worldwide Interoperability for Microwave Access) in order to resolve problems such as frequent topology partitioning due to vehicle high mobility, disconnections in long range communications, and broadcasting in bad weather conditions and high traffic density scenarios. Currently, VANET initiatives and projects highlighting the benefits of vehicle-to-infrastructure (V2I), and infrastructure-to-vehicle (I2V) connectivity to improve monitoring and reporting of road conditions to vehicles have been deployed around the world with success but not too much speed of adoption. Between 2011 and 2013, Telsfeld Telematik and the Car2Car Communications Consortium conducted a pilot study to test V2V and V2I technologies in the European public roads demoing features such as "Green Light Optimal Speed Advisory,"[1] "Green Wave Speed

---

[1]A Green Light Optimal Speed Advisory (GLOSA) system suggests speeds to vehicles, allowing them to pass through an intersection during the green interval. This speed is assumed to be constant until the beginning of the green interval and sent as advice to the vehicle. "Characterising Green Light Optimal Speed Advisory trajectories for platoon-based optimisation," Authors: Simon Stebbin, Mark Hickman, Jiwon Kima, Hai L. Vu, Transportation Research Part C: Emerging Technologies, Volume 82, September 2017, pages 43–62, Elsevier

Information,"[2] "Remaining Phase Time Display,"[3] as well as automated warnings for hazardous locations, road works, weather warnings, and park and ride information. In 2012 the "Tomorrow's Elastic Adaptive Mobility (TEAM)," coordinated by Fraunhofer[4] and with the participation of BMW, VOLVO, NEC, TNO, INTEL, HERE, and others, and additional funding of 17 million euros from the European Union, carried out field and integration tests across various European locations: Tampere (Finland), Gothenburg (Sweden), Berlin (Germany), Turin and Trento (Italy), and Athens and Trikala in Greece. The pan-European project aimed to demonstrate how proactive urban and inter-urban monitoring and co-modal route planning and smart intersection management for buses could optimize public transport for municipalities. The technologies deployed

---

[2]Green Wave Speed Guidance for Signalized Highway Traffic allows in-vehicle systems to provide the driver with speed advises through the variable message signs and the use of an in-car display so the vehicle can keep its cruising speed without stopping at red lights. The system knows the time phases of each traffic light and calculates the correct speed that the vehicle should travel to always go through green lights. The alternative is Eco-Driving Speed Guidance Strategy (EDSGS), which calculates the best travelling speed for a vehicle to save the most energy and power. The eco-driving vehicle's velocity trajectory is smoother than that of the green wave vehicle, and the average compliance rate of EDSGS is higher than GWSGS. The EDSGS showed more benefits than the GWSGS. Compared with the vehicles without speed guidance, the fuel consumption and $CO_2$ emissions can be reduced by 25% and 13% under the EDSGS and GWSGS, respectively. "Eco-Driving Versus Green Wave Speed Guidance for Signalized Highway Traffic: A Multi-Vehicle Driving Simulator Study," November 2013. Procedia – Social and Behavioral Sciences 96:1079–1090. DOI: 10.1016/ Authors: Dening Niu and Jian Sun. Tongji University

[3]Numerical display of the remaining time on a traffic light before it switches to another color.

[4]Fraunhofer, headquartered in Germany, is the world's leading applied-research organization with a focus on developing key future technologies and their commercialization by industry and governments. The majority of the organization's 28,000 employees are qualified scientists and engineers, who work with an annual research budget of 2.8 billion euros. Of this sum, 2.3 billion euros is generated through contract research.

involved a functional Cloud system that coordinated all V2X infrastructure communications using the IEEE 802.11p standard, the data from embedded systems in vehicles transmitting via 4G Lite networks and the involvement of pedestrian smartphones to detect humans on the move.

In 2015, at Amsterdam Intertraffic,[5] an international leading conference and exhibition for the traffic and mobility industry, one of the most discussed issues was to define the roadmap between automotive industry and infrastructure organizations for the initial deployment of Cooperative ITS in Europe. In specific, the importance of standardizing SPaT[6] and MAT[7] communications, which are critical vehicle communications at traffic intersections, the spot where the majority of traffic accidents occur because of their complexity in handling crosstown traffic. The way MAT comms work is by understanding without errors the topological definition of lanes within an intersection and the topological definition of lanes for a road-segment. They also handle the signalling, demarking the links between the various road segments such as each type of lane (speed and slow lanes for merging onto traffic or leaving a highway), and most importantly, the lanes that are restricted or closed. In SPaT communications, the most relevant messages are those related to signalling the phase of traffic lights switchings and the timing of vehicles in and out of an intersection, the status of traffic and the prediction of data elements for prioritization response. SPaTs are meant to correctly issue clear instructions for abstract permissions (accelerate or slow down at

---

[5]Taking place since 1972, this international exhibition aims to develop awareness and collaborations among the fields of infrastructure, traffic management, safety, parking, and smart mobility.

[6]A Signal Phase and Timing (SPaT) message defines the current intersection signal light phases. The current state of all lanes at the intersection are provided, as well as any active pre-emption or priority.

[7]MATs are Meridian Administration Tools in telecommunications for maintaining and updating the Meridian system of telephone exchange switching systems, the first telecommunications infrastructure to allow international calls and data exchanges.

the amber phase of traffic lights?) and for vehicle maneuvers across lanes. Creating international standards for SPaT and MAT communications ensures that all connected objects interpret reality in the same way, that the data tagging of each action and signal is the same in every country, and that every vendor assures their adherence to the standards in order to obtain certifications for commercial deployment. Germany, determined to roll out smart traffic services, launched the Cooperative ITS Corridor, a project supporting all environments – urban, rural, inter-urban, and all V2I2V and V2V communications. C-ITS was phased out across the various states in the country from 2013 to 2018, demonstrating the simplicity and affordability of the infrastructure deployment and the increased benefits that these systems could provide for the traffic population that local authorities could finance with moderate budgets.

The success of the European projects soon encouraged some municipalities in the United States to carry out their own pilot projects. One of the early ones took place in 2015 on three locations of Interstate 80 in the Wyoming area. This is a highly congested corridor for freight and passenger traffic that lacks alternative routes, forcing the more than 32 million tons of yearly freight deliveries to run through it in Wyoming's extreme weather conditions that range from snow blizzards to dense fog in the winter months. Around seventy-five roadside units using advanced dedicated short-range communication (DSRC[8]) technology were installed along identified hotspots in sections of Interstate 80. These RSUs allowed 400 instrumented fleet vehicles equipped with DSRC-connected on-board units to broadcast basic safety messages, share alerts and advisories, and collect environmental data through mobile weather sensors. You may rightly ask yourself if truckers do not already do this and the answer is that

---

[8]Dedicated short range communication (DSRC) refers to two-way radio communication operating on the 5.9GHz band for the purpose of supporting vehicle to vehicle (V2V) and vehicle to infrastructure (V2I) traffic applications. The FCC has set aside this band for this purpose.

they have been doing it for years via Citizens Band radio (CB radio), the famous trucker communications radio frequency used to find out about nearby accidents, roadblocks, weigh stations, speed traps, and traffic, but the point now is that we are making their vehicles take care of that as an automated service not dependent on human involvement. The WYDOT Traveler Information in the Wyoming project delivered real-time, live travel data collected by fleets and roadside units and was made available via a mobile application, *Wyoming 511*, and a commercial vehicle operator Internet portal (CVOP). Other states soon saw the value of this for their road safety programs and, in 2017, the National Operations Center of Excellence launched the *SPat Challenge*, a project to incentivize state and local public sector transportation infrastructure owners and operators to cooperate together to achieve deployment of DSRC[9] infrastructure with SPaT and MAT broadcasts. The challenge aimed to deploy approximately 20 signalized intersections in each of the 50 states by January 2020. Currently, the uptake has only managed to achieve 17 operational DSRC intersections and around additional 35 locations planned across 26 states. Though the Covid-19 pandemic has halted much of the infrastructure activities, the project seems relatively on course to be achieved if not in 2020 as planned, at least before 2025.

Awarding, signing, and building infrastructure of this kind takes time because the telecommunication vendors need to prove the success and return on investment of each case study put on the negotiating table. Additionally, some states are more technology-oriented than others, and thirdly, it is all about local economies. Denver in Colorado is one

---

[9]The National Highway Traffic Safety Administration (NHTSA) is in the process of requiring all new light vehicles sold in the United States to be equipped with DSRC radios and for those radios to transmit basic information about the location, speed, and critical operations of the vehicle. This will enable agencies to both collect limited anonymous vehicle data using roadside installed DSRC radios and to transmit data, such as SPaT, back into the vehicle with the intent to support safer, more efficient operations.

of the most active states because of their digital vision and funding abilities. Some of their recent projects have turned highways into "smart pavements" by installing sensors wedged within the concrete slabs of road intersections to measure the speed, weight, and trajectory of vehicles that pass over it, features that are still very low in automated signalling for V2X but that start proving that this information is useful to the county. The U.S. Department of Transportation (USDOT) is keen to spread the use of DSRCs, but they are still aware that more successful case studies must be brought to market in partnership with both industry and municipalities.

Knowing that the spirit of competition is alive among cities that want to create their footprint in the world within this nascent technology, they launched in 2015 the *America's Smart City Challenge*, getting an overwhelming response from 78 mid-sized cities in the country who presented their urban needs and road management problems and proposed novel ideals to solve them via smart transportation systems. Of the seven finalists,[10] a great majority proposed DSCR solutions and building a data analytics platform that would derive ways to implement traffic management and allow the authorities to make better decisions. More than half of the applicants were keen to test automated share-use vehicles to safely transport individuals to their city destinations. This is the current project being deployed in Austin, Texas in partnership with Google Waymo. Other proposed approaches included installing inductive wireless charging for electric vehicles, fitting public spaces and city transportation with free WiFi, and for those cities with considerable freight deliveries, they proposed implementing smarter curb space management (through sensors, dynamic reservations, and other technologies) to speed loading and unloading. Above and beyond, what emerged clear to the USDOT is that these seven finalists were committed to roll out diverse and inclusive solutions for all citizens, reaching underserved communities, and increasing access to jobs. They were mandated to "reinvent"

---

[10]Austin, Columbus, Denver, Kansas City, Pittsburgh, Portland, and San Francisco.

transportation and redefine what the concept of "smart" should mean when improving the quality and reliability of their transit services and to deliver more affordable and sustainable mobility services to their citizens. The winning city was Columbus, Ohio, awarded with a $40 million prize. The project that the city proposed was to deploy a connected vehicle environment along seven major corridors, including 16 of the top 100 high-crash intersections in the Columbus area.

The 2020 decade is blooming with projects across all regions. As of December 9, 2020, *SmartColumbus* has deployed *Pivot*, a mobile application that delivers real-time transport information and turn-by-turn instructions for the region, also promoting electric vehicles and ride sharing services that are educating citizens on emerging transport solutions as well as creating opportunities for vendors to test their products and services in real-life environments. All other finalists have continued with their own proposals with other sources of funding, determined to become model cities for the 2030 Smart Society, which seems to continue gaining traction across the developed world. In Europe, the place where the first pilots began, the projects now encompass wider areas across borders. The European Union has recently funded a project called MAtchUP to deploy large-scale, live Big Data systems that monitor transportation and traffic conditions in medium-sized cities[11] via digital platforms managed by their local authorities.

---

[11]Valencia has been declared the first Spanish City 100% Smart, thanks to "Vlci" its digital Platform integrating the main local services. The results from the project showed that this ICT tool boosted the quality of life of its citizens, cut the city expenses and improved its efficiency. MAtchUP's other cities are Dresden (Germany) and Antalya (Turkey). The project will also support the replication and upscaling plans in Ostend (Belgium), Herzliya (Israel), Skopje (North Macedonia), and Kerava (Finland). The MAtchUP project is one of several EU-funded projects focusing on smart cities and communities.

# e-Mobility Around the World

All in all, as of February 2020, over 47 cities across the world were piloting self-driving cars, while others were focusing on rolling out autonomous public transport. These initiatives are converging the efforts of both governments and private sector companies in achieving smart urban living objectives, resulting in infrastructure provisioning contracts that are likely to alter and change the shape and rhythm of cities, the type of people that live in them, and the activities that are likely to become ways of life in this new decade.

In specific to the automotive sector, it is also evolving the role and business model of vehicle manufacturers. Many tier 1 automakers have created research and development partnerships with governments and city municipalities to explore the future of smart mobility. The Qatar Investment Authority (QIA) is preparing to deliver their first autonomous public transport in time for the next FIFA Club World Cup in 2022. To deliver this vision, it has created a collaborative partnership with Argo AI, a company invested by Ford and Volkswagen Group developing self-driving technologies, and contracted additional resources provided by Scania, a major manufacturer of commercial vehicles – specifically heavy lorries, trucks, and buses – which has also transitioned to self-driving, and MOIA, a fully electric ridepooling Berlin startup acquired by Volkswagen. This consortium of companies is an example of what other cities with other companies are currently deploying to test the concept of "smart mobility" around the world. The roles across the members of the consortium range from vehicle providers, to software and autonomous vehicle management platforms.

Smart mobility does not mean less vehicles on the road, but better management of them. An example of this is being planned for New York City's Lincoln tunnel, an access point into the city's heart from the New Jersey side that has a daily run-through of approximately 1,850 buses and that the Port Authority of both cities want to expand to additional 200 self-driving buses, increasing the commuter capacity by an extra 10,000 people.

The project seeks to prove that self-driving vehicles, capable of moving closer to each other safely thanks to platooning technology, will increase the efficiency of the commuting route by 30%. This basically turns roads into rail tracks, a vision that some senior people in the car industry dislike because, to their minds, this is nothing to do with private vehicles, but with buses and mini shuttles *which are not their core business.* This is why automotive giants like Volkswagen are investing in the periphery of smart transportation, because the future, in addition to autonomous vehicles, is also about intelligent software platforms to manage diverse flows of traffic. They do not intend to run these services as a core business, but to learn from these projects to gain inspiration for the future of their vehicle fleets.

Still, cities are moving cautiously toward fully autonomous public transport. The Lincoln tunnel project is meant to be carried out without real passengers. Safety is something that still concerns the public, especially local authorities.[12] Not only do people feel uneasy aboard self-driving cars, but they also worry about self-driving projects testing autonomous vehicles taking place in their neighborhoods. To mitigate this, the Institute of Electrical and Electronics Engineers (IEE), a professional organization headquartered in the United States and encompassing senior executives across the wide spectrum of the ITC industry, has put together a first attempt to create an industry standard for safety-related automated vehicle behavior. The working group officers come from the relevant players in the field: Intel, Google Waymo, Uber, and the IEEE itself. They are clear in their thinking that this is not an exhaustive guarantee for the complete safety of autonomous driving systems, but an approach to create best practices for the rules of the road within the context of algorithmic programs entrusted with dynamic driving tasks.

The need for this derives from the consensus across governments and all industry stakeholders that only open transparency and the sharing of

---

[12]A January 2020 survey from Deloitte, "Examining Auto's Future," discloses that 48% of US consumers are apprehensive of driverless cars.

information can truly deliver progress and safety objectives in evaluating the performance of automated driving systems and how they are to qualify for certification to operate in the public roads. Furthermore, drivers' licenses to operate Level 4 autonomous vehicles will have to adhere to the standard. In order to ensure diversity and compliance, the standard will establish a formal rules-based *mathematical model* for verification. The standard, intending to be adopted on a global scale, will also be technology-neutral – any vendor will be able to submit their AV systems for certification, and adapted to regional customization by local governments. It will also include a test methodology and the necessary tools to perform verification of an AV and its conformance with the standard. Basically, the rules of the analogue road of years past will be upgraded to digital compliance. At Level 5 of driving autonomy, where AI[13] becomes the main driving brain for a vehicle, the motor industry seems to veer rapidly toward the provision of public transportation of small shuttles, city buses, and goods deliveries rather than private vehicles. In order to achieve certification for this level of autonomy, they will have to deliver increased safety in urban centers, over and above what is needed for roads and freeways, which are steadier and easier to navigate environments.

# Summary

The coherence of these initiatives favor a 2021 Smart Society agenda that aims to build and sustain urban centers that are inclusive, not just across all aspects of human existence and economic power but also of biodiversity – containing natural spaces and habitats for flora and fauna, as well as becoming a dynamic space able to transform itself according to needs and priorities. Currently, 55% of the population lives in urban

---

[13]Due to the growing use of applications based on artificial intelligence in the automotive industry, it is estimated that the value of A.I. in automotive manufacturing and cloud services will exceed $10.73 billion by 2024.

centers that are being converted into smart cities by local authorities that seek to solve very specific problems: transportation, housing, CO2 emissions, and safety. Below this, you find management efficiency of city services – from books lent at public libraries to potholes fixed on the roads. When you have to handle large volumes of population, you need to start looking at efficiency, reliability, and safety issues. Because large cities are places of continuous motion, how we move things within and out of them becomes a major issue for local authorities. It is no longer something that can be postponed. Automation has transformed the way we manufacture goods, and so traffic needs to start evolving from analogue to digital practices. Traffic lights that currently remain neutral to the type of traffic that goes through them will have to start becoming "smart" and prioritize freight movements, and business to business (B2B) mobile applications will have to be developed to address the needs of the trucking sector, providing truckers with information about routes and parking. Inevitably, when you start optimizing city services, other elements affecting the future of city life quality begin to emerge. Climate change and adherence to CO2 emissions requirements are also forcing cities to study the installation of electric vehicle infrastructure. If conscientious citizens start switching their fossil fuel power units for electric solutions, cities have to provide sufficient places to recharge such vehicles on the public streets and roads. Converting public fleets and buses to electric and bio-gas vehicles is currently a conversion that many municipalities are testing, and there is no single solution to the problem, but homogenized approaches. In mega-cities of millions, local authorities are incentivizing shared-use mobility options, and the services of companies such as Uber and Lyft are considered "smart" city solutions, though this is still debatable as an end-game. Their business models work because the price per ride is heavily subsidized by each company and their investors, and above it all, they are still glorified taxis. What needs to improve and be invested in is better public transportation and the enlargement of safe cycling grids, which are assets directly related to how high the quality of life is in an urban center.

Smart is the new attribute, whether we speak of vehicles, cities, services, or people. Now that we are entrusting intelligent machines with the computational workloads of the digital life that we have created, the remaining cognition is guided by emotional intelligence qualities, abstract thinking, creativity without restrictions, intuition, and all other forms of finding solutions to life problems via biological approaches. The future is smart, and humans will be at the center of it.

# CHAPTER 10

# Digital

*Humankind has been storing data since the Bronze Age. We have thrived and succeeded across civilizations and times thanks to our ability to learn from our own experiences and from the teachings of others before us. Our information obsession has made us turn to digital everything that exists, and for those objects that were biological or analogue, we have fitted them with sensors and extracted all the informational value that they always possessed, a treasured commodity that is allowing us to transform not just the automotive industry but an entire periphery of synergistic sectors like insurance and healthcare as well as the future of cities and how they serve and accommodate their dwellers.*

## Life Is Digital

In 2011, Marc Andreessen, a former technology entrepreneur–turned–venture capitalist,[1] wrote an article for *The Wall Street Journal* announcing the inevitable: software was eating the world, and there was no turning back. Andreessen wrote the article at the time when the technology sector was evolving into a transformative force, not just an industry capable of sustaining itself on IPOs. He remarked how leading Silicon Valley companies were making inroads into software-driven ventures such as the

---

[1]At the time, Marc was co-founder and general partner of venture capital firm Andreessen-Horowitz, which has invested in Facebook, Groupon, Skype, Twitter, Zynga, and Foursquare, among others. He was also an angel investor in LinkedIn.

© Inma Martínez 2021
I. Martínez, *The Future of the Automotive Industry*,
https://doi.org/10.1007/978-1-4842-7026-4_10

mobile industry and artificial intelligence. HP had just bought Palm the previous year to ram their old webOS platform onto it, and immediately thereafter ventured into Big Data analytics by acquiring Autonomy, a Cambridge, England, company specialized in the analysis of large-scale unstructured data, using adaptive pattern recognition techniques centered on Bayesian inference, a precursor technique to teach machines to develop understanding of contexts and rules-based scenarios. The corporate acquisitions frenzy was not a typical "fire sale" of companies in distress, but bold moves into amplifying world domination. When Google made the intrepid move to acquire cellphone handset maker Motorola Mobility, it was a clear signal to the industry that they were fully committed to build up assets in the mobile ecosystem, not just further developing their Android platform. In 2014, Google snatched the jewel in the Deep Learning crown by buying DeepMind, a London-based company that, over the previous four years, had amassed the top talent in neural machine learning expertise. Since then, DeepMind has entirely turned Google's business from a reordering links search engine to the most powerful AI company in the world. "Digitization" and "Disruption" became the most searched technology terms on the Internet. I remember being asked on stage at conferences what it all meant for business and the economy and, following Andreessen's metaphor, I would reply that the most interesting question was, beyond the fact that we were building a digital society, to know who was running the restaurant, who was going to control anything and everything in a digital society that was expanding beyond the World Wide Web.

Ten years later, life is officially digital because data is everywhere. We have turned into bytes everything that exists, from music files to images to climate data, navigation, commerce, banking, and ourselves. We have digitized our biology by creating avatars, second lives, and digital profiles that represent our identity when we use digital services. Digitalization of all aspects of life has created a new layer of informational data about humans and their activities that companies in cybersecurity and user profiling are beginning to deconstruct and rebuild into digital fingerprints of our personas, as precise and unique as the ones we have on the tip of our fingers. Digital identity is one of the fastest growing areas of cybersecurity and of government concern, since paper, for both money and passports, will eventually cease to exist as we evolve to create a "proof" society, one where online identity resides in our ability to demonstrate that we are *who* or *what* we say we are during digital transactions and interactions. Digital identity is set to become one of the defining features of the next stage of human digital transformation because of the plethora of human data sources available to governments and companies.

# Driver Data

Customer profiling, on the other hand, is how companies build up data repositories of our digital behavior and the personal data that we have willingly shared with them, such as our email address, zip code for purchase deliveries, and our birth date to prove that we are older than a certain age in order to gain access to a digital service. The trouble is that many companies have abused customer trust and mined their data without their permission, implanting software crawlers into their devices in order to monitor what else these customers did elsewhere, in other websites, or worse, in all their digital interactions. Quite simply, the knowledge of exactly who or what we are dealing with is a prerequisite of all communication and exchange, and yet, in digital environments, it is a cumbersome and insecure context. What represents identity? A date of birth? An email? A postal address? The IP address of a machine? The answer is all that and more.

This hot spot of human data mining is going to get hotter with the arrival of driving data. Because human identification data is contextual and it varies as its purpose varies, traffic behavior data is a very unique data layer to incorporate into the digital profiling of drivers. Traditionally, for every purpose we intended, we were required to produce a specific set of data. Want to know how the planets and the celestial objects were in the skies at the exact time that you were born? Welcome to astrology. All you need to produce is your date, time, and place of birth, and instantly, you will understand your place in the universe. Want to travel to a foreign country? Your government will issue you with a passport, a set of informational data that has been conferred to you by the authorities in order to single you out from all other citizens in the world. Need access to a service? Your email address and a secret password of your choice will do. Need to demonstrate that you are the rightful holder of a bank account? Input all informational data sets you are asked and you will be allowed to execute transactions. How we prove our identity in one digital context is almost never how we prove it in another because we are living the clumsy days of using and re-using variations of easy-to-remember passwords, taking reliability of people and digital personae at face value and this has led to endemic cybersecurity breaches.

With Internet-connected objects, our digital fingerprints, that is, how we use digital products and services will create unique data sets attributable to us and not someone else because digital Identity is a social construct, not a technical battleground. As digital identity becomes more embedded in our lives, some of the socio-cultural aspects of identity will likely influence how we build our identity data sets around how we hope to be recognized, how others choose to see us and how we elect to represent ourselves in different ways for different situations. The relevance of social contexts are as important in the digital world as they are in the physical reality. Digital identity data needs not be as constrained as paper identity documents, and as such they are likely to grow and expand in profiling data in order to acquire wider social significance.

Piecing together the identity data puzzle is what telematics companies in the driver behavior intelligence sector provide for user-based insurance, a modality of car insurance that some insurers are willing to issue at a discount for good driving behavior. What this has created is a more accurate insurance sector when managing driver-centric risk. For years, women paid less in car insurance because we are safer drivers, while young drivers pay the highest insurance premiums for being considered novice drivers. New gender-equality laws have ceased the aforementioned industry practice of safety attribution to genders, but driver event analytics – how a driver handles a vehicle in regard to acceleration, braking, turning, speeding, flowing with traffic, even detecting driver distractions – has not only allowed many people to lower their car insurance costs but also created a new model of value exchange for an industry that used to assess risks based on passive data – gender, age, car model, years of active driver's license. When drivers allow a company to monitor their every move in traffic for financial compensation, the data sets become accurate data and not assumed behaviors. Overtime, the amount of behavioral data monitored allowed telematics companies to build driver profiles and rank their driving style in every possible context, drawing up new user experience scenarios and potentially evaluating and predicting changes in their lifestyle. Some of these solutions are built into mobile applications that leverage handset features such as accelerometers in order to build another layer of data monitoring services, for example, crash detection and analysis. Typical collisions into other vehicles or pedestrians create specific physics in a vehicle that an accelerometer will interpret under diverse formats: side or rear crashes, impacts into stationary objects like a wall, and vehicles falling down a ravine. These assessments are based on how crash dynamics vary according to the velocity and direction of the other vehicle and the physical forces that are generated on impact. Sentiently, the original "freedom of the road" that automakers sold to customers in the twentieth century is diminishing and evolving toward increased safety values and monetary compensations encouraged in the new digital society that monitors everything that happens.

This emerging trend will give way to the birth of neural intelligence in vehicles, a type of learning that optimizes itself according to how a vehicle is used and what it experiences. Neural AI vehicles will be smarter than others of the same year, manufacturer, and model, because they will be programmed with true evolutive AI that will learn from its own driver, develop an adequate understanding of and draw appropriate conclusions from what it experiences. Your driving behavioral data will help train the algorithms of future vehicle intelligence and, in turn, your car will personalize your experience, becoming yet another behavioral data layer of your digital identity. This is how 1990s AI systems using collaborative filters gave way to suggestive personalization, the key success factor of Amazon in its early days. If your car learns your driving habits, your usual routes to regular destinations, the typical gas stations where you fill up your tank, even your most minute tics when driving, the driving experience will become a bespoke environment. A car manufacturer could learn, for example, at what level of tank fuel consumption most customers tend to refill it. If your data, stripped from your personal identificators, merges onto a general data lake of driving behaviors, the whole sector will start optimizing the driving experience based on what people do, rather than what product designers assume that people want.

# Adapting Urban Planning to the New Mobility Paradigm

Behavioral data will also expand to how mobility solutions deployed in cities take place and are adopted as a way of life. In this new decade we assume that mobility data will encompass a fundamental layer of information comprising management systems of traffic, street lights, congregation and dispersion of population in public areas, delivery of goods into buildings, disposal of waste, air quality and noise reduction controls, and other urban and road systems of multiple data streams.

Basically, anything and everything that moves along streets and roads will provide valuable insights and create a common data knowledge base. Each city will create its own profile, based on how their citizens behave within these digital environments and the make-up of each municipality, the local customs and culture, the human factors that define life in each town. Modern cities in the 2020 decade are moveable environments where many data flows converge, both biological and artificial. This data needs to be equally prioritized, scheduled, and allowed to roam free, but this does not mean that we are oblivious to its identification. This is a new layer of data that begins to be analyzed when planning urban centers and designing mobility solutions.

While villages were built to harbor human homes and towns were formed when businesses were incorporated into day to day activities, cities were born out of transportation activities. Cities became goods delivery hubs, where non-agricultural jobs were on offer, where the administration of justice, civil laws, and medical care were accessible, where the leisure and the entertainment venues sold escapism. People had to find places to dwell and make a life out of urban environments, find love, send their children to educational facilities, adopt a city identity, and eventually, carve their destinies. Cities became amalgamations of activities, urban jungles where a multitude of experiences concurred each day from sunrise to sunset and well into the night. The twenty-first-century society is addressing the evolution of urban centers with mobility solutions that go beyond traffic management but that seek the improvement of life quality as we understand it in this new decade.

According to *The Smart City Index 2020*, a report by Swiss business school IMD that ranks worldwide cities according to economic and technological data as well as to what their citizens perceive that their cities must deliver to them, the concerns ranged from mobility to social inclusivity needs. In Boston, ranked number 33 out of 109 cities surveyed, Bostonians believe that their city should provide (in order of priority): affordable housing, solutions to road congestions, access to public transport – all three dealing with habitation and movement; school education, public safety, health services – living conditions and survival; air pollution controls, recycling, green spaces – health and sustainability; reduction of unemployment, citizen engagement, social mobility, leisure amenities, and even anti-corruption efforts – social concerns. In Denver, ranked at number 36, air pollution is the third most valued over public transportation, which falls to position number 5. Los Angeles, ranked at 26, where health services and security take number 3 and 4 positions, public transport drops to number 9. For New Yorkers, at number 10, security is the second most important concern and road congestion drops to number 6. What this ranking highlights is that, in spite of each local idiosyncrasy, mobility and urban planning concern most people over and above other issues.

This is because in 2020 we are not pushing people out of cities, but are building self-sustained, micro-communities within them. Have you noticed how the big shopping centers and supermarkets tend to be outside of the perimeter of cities but not so much the "services" – education, administration, financial, legal – and the hospitals? This was the result of increased consumerism vs. "other necessities." While in the 2000s e-commerce put small city shops out of business because of the high cost of urban commercial real estate, recent gentrification of urban centers are looking to create human-centric environments that demonstrate the need for high streets to be populated again by small businesses. This is now being encouraged and subsidized by local authorities as a way to attract people to develop a community culture, sit in cafes, have real face to face human contact, "regain" the streets from traffic turning them into pedestrian areas. In Eindhoven, a city in the Netherlands who has consistently built a pioneering reputation for smart mobility, the human-centric approach to innovation has encouraged the local authorities to consult with the citizens in the development of smart solutions and to define their own local interpretation of the "Smart Society." The "New Smart" redirects the development of technological innovations toward creating products and services that address real human needs with adequate approaches that enhance the user experience. By collecting social issues from the bottom up where citizens indicate their frustrations, worries, or dreams, city authorities and their infrastructure partners can co-create conceptual solutions to these problems and test them in a real situation or living lab. One of these projects, the conversion of street lights into smart urban lighting allowed them to understand that women feel more apprehensive and less safe in poorly lit streets, whereas certain low level areas of the city favored by runners should be lit by motion sensors illuminating the pathways as needed.[2] What the sensor-based society creates is an environment of informational values of anything connected to a network and automotives play an important role in this development.

When data gathering sensors were incorporated into urban centers and not just to rockets, satellites, or industrial robots, the first readings provided weather information. Pachube (pronounced "patch bay" as in the hardware device used to route the inputs and outputs of audio signals) was a hobbyist's network hosting live data from weather stations that people would hang on their balconies and window sills. It was the first international Internet of Things network and was run like an open source community. Then, the Japan earthquake and nuclear reactor disaster of 2011 happened. For the first couple of days, Japan's Ministry of Education, Culture, Sports, Science and Technology issued periodic updates on radiation levels, but their Geiger readers, hanging about 2–3 meters above

---

[2]The "Your Light on 040" project connects the public lighting in the city to a smart light grid with applications that improve the quality of life while also targeting energy reduction. Another project, the "Stratumseind 2.0 Living Lab" intends to address the poor safety in Stratumseind, one of Eindhoven's most prominent nightlife areas, attracting well over 20,000 visitors on weekend nights. Almost deserted during the day, the area is in decline and the number of severe public incidents on the rise, which the city plans to mitigate via innovative solutions involving lighting, social media, gaming technology, and the collection and processing of sensor data in order to determine the effects of measures and to study which factors contribute to violence and discomfort.

ground, provided a partial interpretation of what the actual radiation could be below such height, at the levels where people, children, and pets lived. The team running Pachube, together with Libelium, a sensor manufacturer based in Saragossa, Spain, sent around one hundred Geiger devices so that the people in Fukushima could measure live radiation data at ground level. The crowd-sourced data was not able to be considered "official data" of the radiation environment, but it proved the importance of something that the government data did not provide: live feeds. It also created international awareness of why Internet-connected devices could be of vital importance in urban environments.

The data that cities are today collecting in order to analyze how we move is helping urban planners to redesign public spaces, especially the available space that removing analogue traffic signals will leave behind when full vehicle automation is achieved and we lift them off streets and pavements. There is a growing concern that the streets will have to give priority to pedestrians in ways that will alter the traffic paradigms of the past. Current cycle lanes created in most European cities are still a dangerous moving flow of sizable objects hard to sort for pedestrians, especially when some cyclists switch from roads to pavements at will. Cyclists in Amsterdam move through the streets at supersonic speeds and through invisible lanes that only the locals know how to avoid. In Berlin, cyclists ride the long stretches of cycle lanes as fast as or faster than the cars, pedalling their bikes right next to pedestrians. God forbid that you take a centimeter of their assigned lane, as a storm of ringers will deafen your ears to remind you that, in the kingdom of city mobility, pedestrians are last.

The European Union's Mobility and Transport organization is keen to remove traffic signals in the hope that they will decrease bad driving habits such as running over amber lights, and going over the street speed limits, but if we well and truly want humans to "recover" the streets, city traffic speeds will have to seriously decrease and not all humans ride bicycles. Many of them are vulnerable people, like the elderly, or have mobility impediments, which is why all pedestrians and their smartphones will become part of what cities are to consider mobility data, simply because we cannot separate flows of movement if we are to create seamless environments of interaction that are inclusive.

# Driver Biometrics: The New Health Data Layer

Human-centric data is also increasingly transforming car interiors. As current car models occupy themselves with how safely we drive and how they can predict drivers' mistakes, they are also turning themselves into health monitoring systems, creating a crossroads between the automotive and the healthcare industries, a "first" which many did not see coming because they doubted that sensor-based technologies and edge computing would be able to create a data gathering and analytics environment that both passengers and doctors would welcome one day. Before Covid-19 not many people knew what an oximeter was for. Today, embedded as a feature in the new Apple iWatch 6, the percentage of oxygen levels in our blood stream reveals our oxygen saturation and peripheral perfusion, the delivery of blood to tissues, such as our lungs' airways.[3] It is an extremely sophisticated biometric, something that only a few used to monitor, but that a global pandemic has turned into a vital signal to check daily.

Health monitoring applications made available to individuals have created awareness of the importance of self-gathering health data to prevent disease. Unfortunately, this is one of the biggest hurdles in the current meditech sector, as there are no standards or industry regulations that force companies to allow customers to download and share at will

---

[3]Covid-19 can be a silent killer. You may breathe with normality and then, when your lungs are running at just 20% capacity, it is when you will notice that, suddenly, you can barely draw oxygen to your bloodstream.

their own medical data gathered and analyzed by third parties offering monitoring products. It is a closed garden of data silos that forces individuals to have disconnected and fragmented visions of their overall health parameters.

Vehicle manufacturers are sitting on a golden opportunity to be the first ones that open up the barriers, well positioned to build alliances with healthcare organizations and medical practitioners because vehicles today can do much more than taking vital signs readings: some of them have been analyzing data that could determine deteriorating mental illnesses. The question for some car manufacturers who are currently developing such health systems within their vehicle seats is no longer if they can interpret the data correctly but if passengers will be willing to know, for example, that they are showing symptoms of Alzheimer's. In conversation with some stakeholders, they let me know that most people would not want to discover such things from their cars right now, but that, eventually, if cars were able to predict health evolutionary paths, drivers would most definitely welcome that their vehicles would alert them that, in a given period of time in the future, they could be susceptible to develop a certain disease or suffer signs of sensorial deterioration and so, be able to work with their health practitioners toward a preventative therapy.

The driver and passengers' experience is veering toward a context where your vehicle looks after your well-being beyond the context of road accidents. Passive health monitoring is not just the realm of iWatches or Fitbit wrist bands. Biosensors are the cornerstone of the future of patient care as the role of medical doctors switches from diagnosing toward creating personalized therapies thanks to the increase of accurate patient data at their disposal. This data will be gathered not just during critical moments of acute situations such as early warning signs of heart or asthma attacks but will also be data related to the long-term issues of prolonged, undetected diseases, data that often times patients are not willing to gather on a daily basis, because they forget about it or consider it unimportant.

Worse, most of the time people are simply unable to gather it because they do not have appropriate biosensor instruments.

In the very near future, a future that is just around the corner, sitting in a car for two hours of work commute each day will allow car seats to gather such daily health data sets without our involvement. If shared with doctors, medical practitioners will be able to perform remote diagnosis, or if fed into a diagnostics device trained with artificial intelligence to detect anomalies and measure the correct functioning of human vital signs, humans will not even need to make medical appointments for health check-ups because their health will be monitored on a daily basis from the comfort of their cars. Automobile manufacturers like Nissan, Ford, and Volkswagen are said to be the key vendors in the global automotive biosensors market. These devices, designed primarily to monitor the health parameters of the driver, have also shown other uses to optimize driving safety. Biosensors monitoring the driver's facial gestures and eye movements can not only detect a variety of emotional states and health warnings but also if the driver is falling asleep at the wheel.

The need to curb road fatalities during night hours is a top priority in the domain of public safety. According to a report published by the National Highway Traffic Safety Administration in 2015, fatality rates are three times higher in the night or during the dark hours for all occupants of a vehicle, not just the driver. Automotive OEMs and Tier 1 car manufacturers have responded to this challenge with technological features that combine imaging sensors with light detection and ranging or laser imaging detection and ranging (LiDAR) technologies, creating biosensors capable of acting as Breathalyzers and eye alertness tracking sensors. It is a vertical that is attracting medical equipment manufacturers like Medtronic, the leader in heart monitoring ECGs and Holters, and Freer Logic, a provider of neurotechnology. Both companies have adapted their health monitoring systems to the vehicle interiors, with Medtronics fitting ECG sensors to safety belts and Freer Logic creating for the automotive industry a neurobiomonitor headrest that provides real-time brain data. Without any direct contact with the vehicle's driver, the neurobiomonitor is able to determine fatigue, drowsiness, cognitive load, distraction, emotion, stress, and relaxation. It has also been programmed to communicate with the driver via voice alerts, to suggest actions such as stopping to rest, or to activate the entertainment system and play some calming music. No, it does not read your thoughts, at least not yet, but it does interpret your brain electrical activity and the areas where it happens, which, in neurophysiology, has allowed scientists to associate it to emotional states. It is a new dimension of human and machine interaction whereby the machine will instruct us to do certain things or activate other systems for our well-being and safety. It is a reverse context to the voice-command systems that we currently use, a familiar scenario where we tell machines what to do. In this approaching future of autonomous systems, our cars will instruct us to obey them for the sake of our safety.

Will we be able to overrun these systems at will? If we follow instructions from a machine designed by a manufacturer, what will happen to us if we do not? Will our vehicles take over and drive us to the nearest resting area, refusing to start until they consider that we are in an acceptable state to continue driving? Until now, no one had considered allowing autonomous systems to override our civil liberties based on physiological data, but what if telematic companies, who currently reduce our car insurance costs when we drive respecting the road code, incorporated vehicle biomonitoring systems to their mix and judged us by them?

Perhaps, in ten years, all autonomous cars will come to market with the ability to override the driver for the sake of preventing accidents, and we will have to accept this as part of the future society, one where the well-being of everyone concerned stands above my individual privileges. We already do this when we halt our vehicles at a stop sign or when we encounter a train crossing, and remain stationary, engine switched off, even though no train is passing and there is no one in sight. We do this because we are taught to respect these signs for our safety. In this future society, and thanks to data gathering sensors, we will trust many of the machines that monitor our health to the point of considering them "the voice of reason." We will accept that their duty to look after our vital signs will be in place for the sake of our own good and for our own merit since many aspects of healthcare provided by governments are still unaddressed and unserviced.

Mental health is precisely one of the most neglected human maladies. It is considered the twenty-first-century cancer for its exponential increase in the last 20 years, and predicted to augment its effect on the worldwide population across all ages and genres. Even though depression continues to increase year on year among young people and adults, a large number of the population continues to be undiagnosed or provided with inadequate therapies and treatments. In a report by Mental Health America entitled "The State Of Mental Health In America," the statistics are concerning: in 2017–2018, 19% of adults experienced a mental illness, an increase of 1.5 million people over the previous year's data sets. The data for 2017–2018 showed that 60% of youth with major depression did not receive any mental health treatment, even in states with the greatest access, where over 38% of youth did not receive the mental health services they needed. Among youth with severe depression, only 27.3% received consistent treatment whereas 23.6% of adults with a mental illness reported an unmet need for treatment, a number that has not declined since 2011. All these statistics are pre-Covid-19. If anxiety, panic attacks, depression, and stress are not dealt with, the automotive industry will have increased barriers to continue ensuring road safety, mainly because drivers will be less fit to drive, which is why most car manufacturers welcome the launch of autonomous vehicles fitted with healthcare biosensors and programmed to help us navigate destabilizing circumstances such as traffic jams or excessive loudness in high traffic areas.

Biosensored cars are being taught to predict and bypass such stressful circumstances and help us make other navigational choices. Car cabins are being insulated against high-pitched decibels and strenuous sounds from the exterior that can affect our mood and emotions, cause irritation and anger, and destabilize our mental states when driving. Other data gathering sensors will measure oxygen levels, indoor and outdoor pollution rates, and detect whatever floats in the air, even pathogens and toxic substances like odorless gases. As cities grow in size, and urban centers are predicted to host 60% of the world population by 2050, the quality of the air we breathe and other factors such as sun radiation, extreme weather conditions, and percentage of pollen particles in the air will be measured and assessed by in-cabin sensors which will be able to design travel routes that will steer away from these hazards.

## Summary

Software ate the world ten years ago, a parable that Andreessen used in order to make companies aware that, if they did not transform themselves digitally, their core businesses would go extinct as their competitiveness would diminish. In the mobility society, the concept of "going digital" escalates beyond the automotive realm into the heart of human life. It is our duty as citizens of the digital society to ensure that governments and the private sector treat our human data with the utmost ethical practices. Furthermore, that when we accept to be monitored, we do so aware of the common good, but we will not tolerate civil liberties infringements, or the abuse of control by companies that feel entitled to treat us as data commodities. Much is to be gained with the digital revolution of urban centers, roads, and transportation. Even greater will be the benefits to healthcare and lifestyle services. All that we ask, as people, is to remain valued, respected, and cared for.

# Index

## A

Absolute innovations, 16
Acura sound system, 28
Adaptive Cruise Control
    (ACC), 52, 125
Advanced Mobile Phone System
    (AMPS), 45
AI natural language systems, 31
America's Smart City
    Challenge, 169
Apple-compatible car systems, 35
Automakers, 107–109
Automated parking, 52
Automate Lane Keeping System
    (ALKS), 128
Automation, 97
    AI, 15
    benzos, 14
    BMW, 14
    car manufacturing, 14
    data car electronics, 15
    data hypotheses/
        assumptions, 15
    engine management sensors, 16
    ESC, 14, 15
    FORD, 14
    German Autobahns, 14
    GM vehicles, 15
    Mercedes-Benz, 14, 16, 17
    smart key, 16
    Stabilitrak, 15
Autonomous Land Driven Vehicle
    (ALV), 120
Autonomous public transport, 171
Autonomous vehicles, 27, 98, 172,
    173, 191
Autonomy
    AI-based
        ALV, 120
        approaches, 118
        ARGO Project, 121
        automated highway
          systems, 120
        autonomous innovations,
          119
        autonomous
          vehicles, 117, 119
        budgetary restrictions, 120
        Captcha, 118
        CMU NavLab, 119
        DARPA, 117
        data, 118
        EUREKA, 121
        German Autobahns, 119

© Inma Martínez 2021
I. Martínez, *The Future of the Automotive Industry*,
https://doi.org/10.1007/978-1-4842-7026-4

# E

Printed in the United States
by Baker & Taylor Publisher Services